Wiley Biotechnology Series

Series Editors: Professor J. A. Bryant, *Department of Biological Sciences, Exeter University, UK*, and Professor J. F. Kennedy, *Department of Chemistry, University of Birmingham, UK*.

This series is designed to give undergraduates and practising scientists access to the many related disciplines in this fast developing area. It provides understanding both of the basic principles and of the industrial applications of biotechnology. By covering individual subjects in separate volumes a thorough and straightforward introduction to each field is provided for people of differing backgrounds.

Published Titles

Biotechnology: The Biological Principles M. D. Trevan, S. Boffey, K. H. Goulding and P. Stanbury

Fermentation Kinetics and Modelling C. G. Sinclair and B. Kristiansen (Ed. J. D. Bu'lock)

Enzyme Technology P. Gacesa and J. Hubble

Animal Cell Technology: Principles and Products M. Butler

Fermentation Biotechnology: Principles, Processes and Products O. P. Ward

Genetic Transformation in Plants *R. Walden*

Plant Biotechnology and Agriculture K. Lindsey and M. Jones

Biosensors E. Hall

Biotechnology of Biomass Conversion M. Wayman and S. R. Parekh

Plant Cell and Tissue Culture A. Stafford and G. Warren (Eds)

Biotechnology in the Food Industry M. P. Tombs

Bioelectronics S. Bone and B. Zaba

An Introduction to Fungal Biotechnology M. Wainwright

Biosynthesis of the Major Crop Products P. John

Monitoring Genetically Manipulated Organisms in the Environment C. Edwards (Ed.)

Plant Cell and Tissue Culture

Edited by

Angela Stafford

Plant Science Ltd, Sheffield, UK

and

Graham Warren

Department of Molecular Biology
and Biotechnology,
University of Sheffield, UK

JOHN WILEY & SONS

Chichester · New York · Brisbane · Toronto · Singapore

Other Wiley Editorial Offices

John Wiley & Sons, Inc., 605 Third Avenue,
New York, NY 10158–0012, USA

Jacaranda Wiley Ltd, G.P.O. Box 859, Brisbane,
Queensland 4001, Australia

John Wiley & Sons (Canada) Ltd, 22 Worcester Road,
Rexdale, Ontario M9W 1L1, Canada

John Wiley & Sons (SEA) Pte Ltd, 37 Jalan Pemimpin #05–04,
Block B, Union Industrial Building, Singapore 2057

British Library Cataloguing in Publication data

Plant cell and tissue culture.
  1. Plants
  I. Stafford, A.   II. Warren, G.   III Series 581

ISBN 0 471 93261 2

Typeset by Vision Typesetting, Manchester
Printed in Great Britain by Dotesios Limited,
Trowbridge, Wiltshire

Contents

List of Contributors

Eunice Allan, School of Agriculture, University of Aberdeen, 531 King Street, Aberdeen AB9 1UD.

Ray Cresswell, Plant Science Limited, Firth Court, Western Bank, Sheffield S10 2TN.

Alan Scragg, Molecular Biology and Biotechnology, The University, Sheffield S10 2TN.

Angela Stafford, Plant Science Limited, Firth Court, Western Bank, Sheffield S10 2TN.

Gagik Stepan-Sarkissian, Plant Science Limited, Firth Court, Western Bank, Sheffield S10 2TN.

Graham Warren, Molecular Biology and Biotechnology, The University, Sheffield S10 2TN.

List of Abbreviations

ABA	abscisic acid
AHS	acetohydroxyacid synthase
6BA	6-benzyladenine
BAP	6-benzylaminopurine
bp	base pairs
CaMV	cauliflower mosaic virus
CMS	cytoplasmic male sterility
CNS	central nervous system
ctDNA	chloroplast DNA
2,4-D	2,4-dichlorophenoxyacetic acid
DAPI	diamino-2-phenylindole
DHFR	dihydrofolate reductase
DMDP	dihydroxymethyl-dihydroxypyrrolidine
DMSO	dimethyl sulphoxide
DNJ	desoxynojirimycin
EPSP	5-enopyruvyl shikimate-3-phosphate
FACS	fluorescence-activated cell sorter

GA	gibberellic acid
GC	gas chromatography
GS	glutamine synthetase
hGH	human growth hormone
HPLC	high-performance liquid chromatography
IAA	indole-3-acetic acid
IRS	infrared spectroscopy
MGT	mean generation time
MS	mass spectrometry
MSO	methionine sulphoximine
mtDNA	mitochondrial DNA
NAA	naphthalene acetic acid
NMR	nuclear magnetic resonance
ODC	ornithine decarboxylase
OUR	oxygen uptake rate
PAL	phenylalanine ammonia-lyase
PCV	packed cell volume
PEG	poly(ethylene glycol)
RIA	radioimmunoassay
SCP	single-cell protein
STOX	(S)-tetrahydroprotoberberine oxidase
T-DNA	Ti-plasmid DNA
TIBA	tri-iodobenzoic acid
TLC	thin-layer chromatography

Chapter 1

Plant Cell Culture

EUNICE ALLAN

A Perspective on Plant Cell Culture

In recent years, the *in vitro* cultivation of plant cells has become a relatively simple task. This simplicity can be illustrated using the routine schoolroom experiment which demonstrates callus formation on carrot. The experiment is straightforward in terms of materials (carrot, scalpel, disinfectant, sterile water and growth medium), methods (surface-sterilize carrot, wash, cut a piece of tissue and place it on the growth medium), and results (callus usually forms within a week). Indeed the experiment can even be extended, by providing another growth medium, to illustrate cell differentiation and shoot formation. In the late 1960s such simple methodology acted as a foundation for grandiose speculation. Predictions were made concerning the creation of novel hybrid crops and cell culture systems which would over-produce natural biochemicals. This conjecture, along with its accompanying excitement, has now subsided and in its place lies an enthusiasm for investigation which, it is hoped, will achieve similar aims.

In order to appreciate the achievements in plant cell culture technology, some knowledge of its history is required. Gautheret's (1985) excellent historical review illustrates the development of technical and theoretical aspects which led to the establishment of plant cell culture as an important area of biotechnology. Consideration of the time scale necessary to provide experimental evidence supporting the fundamental hypotheses of plant cell culture will perhaps place current research endeavours in perspective.

The ability of an individual cell to grow and divide in a self-regulating (i.e. autonomous) manner was first expounded in the cell theory of Schlieden (in 1838) and Schwann (in 1839) which also incorporated the concept of totipotency, i.e. that an individual cell can regenerate into a whole organism. These concepts were easily realized for many vegetative cells but were more difficult to prove for those differentiated cells which, within organized tissues, did not grow and divide. Many researchers maintained plant cells *in vitro* but the multiplication of a single cell was not verified until 1954. Indeed the regeneration of a single cell (in this case tobacco) was only accomplished in 1965 by Vasil and Hildebrandt. These achievements were themselves associated with many important discoveries such as the use of synthetic plant growth regulators, protoplast production, and tissue regeneration. Thus, it took almost a century to develop the techniques which are described today as 'simple and routine'. In essence, plant cell culture technology is a recent innovation which has progressed rapidly in the last 30 years and is now recognized as a major discipline playing equally important roles in fundamental biological studies and product-orientated research.

It is hoped that this book will illustrate some of the many important advances that have been made in plant cell and tissue culture, and indicate what course may be taken in the future.

An Introduction to General Techniques and Technology

A preliminary introduction to some general terms and methodologies is appropriate before discussing the advantages of plant cell culture systems.

The essential cell material in most plant cell culture systems is known as *callus*, which is in reality a mass of undifferentiated cells. Initiation of callus is undertaken by removing material, known as the *explant*, from the whole plant. The explant is surface-sterilized and transferred aseptically onto a complex semi-solid medium supplemented with plant growth regulators (usually an auxin and a cytokinin). Subsequently cell proliferation occurs and callus forms. The term 'callus' was originally derived from the whole plant, where similar cell proliferation is a characteristic response to injury. Callus can be removed from the explant and maintained *in vitro* by routine subculture or by various storage techniques. Alternatively the callus can be manipulated to cause cell differentiation into organized tissue and hence whole plants can be regenerated. In addition to these techniques callus can be introduced into similarly complex but liquid medium and agitated so that the cells disperse throughout the liquid to form a cell suspension culture. Such cells are in theory also totipotent and, concomitant with this, they should also have the potential to synthesize any of the compounds normally associated with the whole plant. Figure 1.1 shows a schematic diagram of the procedures used to initiate callus and suspension cultures.

Suspension cultures are composed of autonomous cells which are dispersed evenly throughout the medium. Consequently these cultures are physiologically more homogeneous (and therefore potentially more controllable) than the whole plant and hence offer advantages for the production of plant biochemicals. The advantages of homogeneity and totipotency are exploited in plant cell suspension culture since the cells can be grown in very large volumes, 10,000 L or greater, and manipulated to produce the desired compound. Indeed, the use of plant cell and tissue culture for production of biochemicals under controlled conditions removed

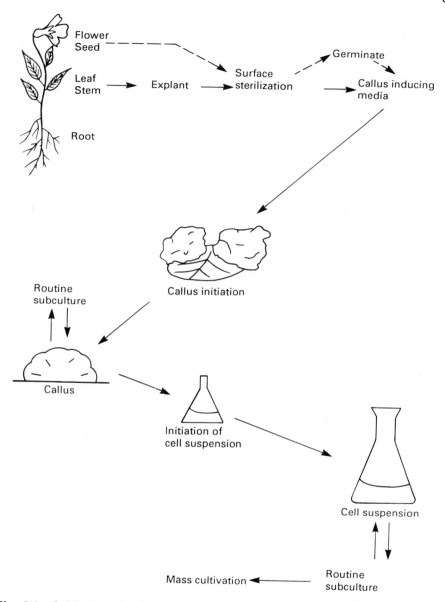

Fig. 1.1   Initiation of callus and suspension cultures.

from the various environmental (and possibly political) constraints of the plant's natural habitat has many advantages. The ability to produce materials of consistent quality in predictable quantities as desired would also be of major benefit to many manufacturers, while perhaps most importantly of all, plant cell and tissue culture could provide a means for increasing product yields far beyond those already obtained in the whole plant (see Chapters 6 and 7). Another advantage of plant cell culture systems is the possibility of production of *novel* biochemicals, not known in the whole plant (see Chapter 6).

Two major areas of plant cell biotechnology which are of commercial importance are micropropagation and production of plant biochemicals including enzymes and secondary metabolites. In propagation, vegetative tissues are induced to proliferate in a manner which overrides the control mechanisms normally exerted by the whole plant. Most gardeners propagate plants of certain species by taking cuttings, and so can derive daughter plants from the parent at a speed which is much greater than that imposed by the natural reproductive cycle. In micropropagation the supplementation of defined growth media with suitable plant growth regulators allows tissue differentiation to occur at natural growing points, e.g. on axillary buds, and also at other sites, e.g. on leaf or callus material. In this way many more progeny can be obtained than with traditional horticultural methods. Moreover, a means of vegetative propagation also becomes available to those plants which, because they have only one growing point, are recalcitrant to asexual division. The uniformity of the population derived via some micropropagation methods can be advantageous not only in plant breeding, for such characteristics as crop yield and morphology, but also for the synthesis of biochemicals. On the other hand, micropropagation via a callus phase can create diversity within the population (somaclonal variation), which can be regarded as a menace or an advantage depending upon the technological application. Similarly, genetic variability generated within the more amenable system of cell suspension culture can be exploited for plant improvement through cell selection (see Chapter 5). The skills of traditional plant breeding, micropropagation, cell selection, and genetic transformation will, when used in conjunction, be of great importance to both horticulture and agriculture. Indeed, micropropagation is already used for the production of many ornamentals and crop plants; for example the multinational company Unilever has spent many years developing propagation methods for oil palm (*Elaeis* sp.) based on somatic embryogenesis.

Plant biochemicals are used as medicinals, agrochemicals, flavours, colourings, and aromas, and this plethora of compounds creates an even bigger demand for the use of *in vitro* cell culture. Indeed, many of these compounds are isolated from specific plants whose growth may be restricted to a limited habitat. In addition, the desired product may only be formed at a particular stage of the growth cycle or may be a transient intermediate in the plant's metabolism. These factors provide difficulties in obtaining plant biochemicals and, combined with chemical extraction and isolation costs, decree that many plant compounds are expensive (over $1000/kg). The potential advantages of plant cell and tissue culture over traditional horticultural methods suggest that it could be used for the production of every plant-derived compound; however, this is not the case. The limitations of current techniques and the very characteristics of plant cells themselves (see below) result in plant cell culture being an expensive process. This in turn suggests that without more fundamental research, the production of plant metabolites will probably be limited to those compounds that cannot be synthesized by other cheaper methods (i.e. chemical and microbial processes) and will include high-cost speciality biochemicals such as pharmaceuticals and food additives.

Having briefly introduced plant cell biotechnology, the remainder of this chapter will illustrate the theory and general methods involved in plant cell cultivation (with some emphasis on secondary metabolite formation). These methods are central to plant cell and tissue culture, and their refinement for specialized techniques will be discussed in the chapters on plant regeneration (Chapter 4) and secondary metabolite formation (Chapter 6).

Callus and Suspension Culture Techniques

THE IMPORTANCE OF STERILITY

The main advantage of exploiting plant cell cultures rather than the whole plant is that control of both the physical and chemical environment can be more easily exerted. In order to achieve this one of the most important criteria for plant cell culture is that aseptic conditions must be maintained. Plant cell culture media are composed of many different compounds (see below) and, with the exception of autotrophic cultures, high carbon concentrations (usually 2–5% w/v sucrose) are used. Such a complex medium supports microbial growth and, since these organisms generally have faster growth rates than plant cells (e.g. with doubling times of the order of minutes rather than days), the growth medium if contaminated will be quickly colonized. Growth of micro-organisms will rapidly alter the supporting medium, and hence controlled conditions are lost. In addition carbohydrate moieties produced by micro-organisms may also directly affect plant cell metabolism (often enhancing secondary metabolism) and may irreversibly alter the characteristics of the plant cell line.

Sterile Technique

Sources of contamination are (a) the explant or culture itself, (b) the growth vessels, (c) the growth media, (d) the environment where growth and/or handling are undertaken and (e) the instruments used in handling tissues. Points (b)–(e) are factors of which microbiologists have long been aware, and techniques for sterilization are well known. One major difference in plant cell culture 'sterile techniques' is that they are carried out solely to safeguard the culture; there is no requirement to protect the operator. Indeed, the use of flow cabinets which filter air from the rear of the cabinet towards the operator offer certain benefits over the more sophisticated laminar flow hoods used in microbiology, in that several plant cell cultures and their associated equipment can be handled without major disturbances of the air flow characteristics. If a flow cabinet is not available it is possible to handle plant cultures in a clean room away from the disturbed air of the normal laboratory. Filter sterilization (through membranes of 0.2 μm pore size) can be used for media sterilization using the wide range of commercial filter apparatus available. However, except where media are prepared in bulk, autoclaving (121°C for 15 min) is generally used, with thermolabile compounds (e.g. zeatin, urea) being filter-sterilized. Dry glassware and heat-stable solid materials can be sterilized using dry heat (160°–180°C) for 3 h, although a double sterilizing period is recommended if regular contamination by endospore-forming bacteria is encountered. In addition to these precautions, care must be taken to sterilize all instruments which will have contact with the sterile components. Thus, for example scalpels, forceps, etc. should be flamed in ethanol (70–95% v/v) and cooled prior to use, and contamination by atmospheric micro-organisms can be reduced by flaming the open orifices of culture vessels before and after making any additions (this is advised even when working in a flow cabinet). When a clean room is available, cell cultures can usually be grown in glass flasks with metal caps, aluminium foil, etc. without the added precaution of foam or cotton wool bungs. If cotton wool is used it is recommended that, especially if frequent handling is anticipated, an outer layer of muslin is used to prevent strands of cotton entering the culture. A cell line is unique, and since there is no cure for contamination, its loss may lead to the abandonment of a particular project. The importance of prevention of contamination must therefore be emphasized. Good general laboratory practice, involving cleaning of equipment and the working area (e.g. swabbing with disinfectants), will provide a sound basis for aseptic techniques.

Sterility Testing

In order to ensure elimination of micro-organisms, sterility testing should be undertaken routinely. Considering the fast growth rate of micro-organisms compared to plant cells, any contamination is usually obvious as clouding of the medium, frothing of cultures, or growth of microbial colonies on the surface of the 'semi-solid' callus growth media. These characteristics are often associated with an 'off smell' noticed on opening cultures, and contamination is easily confirmed by microscopy (using a magnification of at least ×400). In certain situations, e.g. screening of stock cultures, and prior to the inoculation of large volumes of medium, it is advisable to check for *low* levels of contamination. This can be achieved using pre-mixed powders of microbial media which are easy to prepare and can be used in solid or liquid forms. Generally, non-defined media for bacteria, e.g. nutrient agar and malt agar, and Czapek Dox medium for fungi should be used, but occasions may arise (e.g. for the identification of a recurring contaminant) when more selective media and cultural conditions are required (e.g. oatmeal agar for actinomycetes). Surprisingly, plant cell cultures do not appear to be plagued by intracellular contaminants such as mycoplasmas which are a major burden to animal cell culture technologists, but whether this reflects the absence of contaminants or less precise screening methods remains to be seen. For liquid cultures, volumes of the sample to be tested can be aseptically removed from the culture vessel and inoculated on or into the appropriate medium. If large volumes are to be used, e.g. greater than 10 mL (for example when checking a bioreactor prior to inoculation) aseptic filtration through sterile membranes of 0.2 μm porosity will retain the micro-organisms on the filter which can then be incubated on the appropriate solid medium. Smaller volumes (0.5 mL–approx. 10 mL) can be inoculated into liquid medium, while even smaller volumes (50 µL–200 µL) can be directly plated (using a glass spreader) on to the surface of solidified medium. Incubation (typically at 25–30 °C) will result in visible microbial growth usually within a period of 1–3 days. Further isolation, growth, and identification of a particular contaminant will follow general microbial techniques (see for example Stanier *et al.*, 1977).

CHOICE AND PREPARATION OF EXPLANT

Cell cultures can be initiated from both mono- and dicotyledonous plants, although in general the latter form callus more readily. Callus will usually develop from any explant of the whole plant, although the choice of explant will be dependent on the research aims; e.g. biochemical studies may require comparison between callus derived from different plants, organs, or tissues, while for secondary metabolite production the plant species will usually be dictated, and in cases where nuclear cytology is being investigated research can be simplified by using plants with low chromosome numbers (e.g. *Crepis capillaris*, where $2n = 6$).

Generally, for callus initiation the explant material should be healthy and vigorous (e.g. young leaves generally produce callus more easily and more quickly than mature leaves) and it is inadvisable to use tissues from plants which are about to enter dormancy. The use of newly germinated plants can be advantageous for callus initiation, especially if seed can be sterilized and germinated in aseptic conditions (e.g. by growth on a standard medium as described below without supplementation of growth regulators). Callus initiation appears to be independent of explant size, although there seems to be a minimum size below which cell division does not occur. In view of this, sterilization of larger materials can be undertaken with subsequent excision into smaller portions for inoculation onto inducing medium.

Explant Sterilization

After choosing the explant it has to be surface-sterilized if not already sterile, and placed on a callus-inducing medium (see below). The nature of the explant may impose difficulties in handling and sterilization. Stem tissues of woody species are often extremely difficult to handle, and the epidermal and bark layers may have to be excised for successful sterilization. Indeed, some researchers have initiated wound callus in the whole plant and subsequently sterilized this material which hence acts as the explant (Dixon, 1985). Similar problems can be encountered when sterilizing seeds; in some species seed coats may have to be removed completely, while others, e.g. *Cinchona* sp., have to be de-husked. Root tissues, normally in contact with the soil microflora, are particularly difficult to sterilize especially if there are fungal associations (i.e. mycorrhizas) and therefore tend to be poor explants for callus initiation. Leaves are generally the simplest to sterilize, although the high level of fungal spores present in autumn may result in the need for more severe sterilization procedures than those used in other seasons. Large fruits or tubers are usually very easy to handle and often only require flaming in ethanol prior to excision of internal tissue. Some tissues, e.g. waxy leaves, tend to be hydrophobic and the addition of detergent (e.g. 1% Tween) will enhance sterilization. These factors tend to imply that sterilants should be applied at high concentrations for long time-periods, but disinfectants are non-specific and will act on the plant tissue as well as on the associated micro-organisms. The sterilization regime should therefore aim to use low concentrations for the minimum duration. A wide range of disinfectants have been used for explant sterilization e.g. calcium/sodium hypochlorite (1–2% available chlorine), hydrogen peroxide (10–12%), bromine water (1–2%), silver nitrate (1%), mercuric chloride (0.1–1%), and antibiotics. Generally hypochlorite solutions, e.g. commercial domestic disinfectants, are easy to obtain and use, and are most effective. As a general guideline, material should, if necessary, be cleaned of soil debris, etc. by washing in tap water. An initial pre-sterilization in ethanol (5–30 s) followed by 1–2% (available chlorine) sodium hypochlorite (10–15 min) is usually sufficient for most tissues. Enclosed clean vessels that have contained the sterilant for at least 5 min are usually adequate for sterilization procedures, although subsequent washings must be undertaken in sterile containers with strict adherence to aseptic technique. In addition, shaking the material during sterilization will obviously enhance the effectiveness of the process.

After sterilization the explant should be washed several times (× 5) in sterile distilled water to effect complete removal of the sterilant. Edges of exposed tissue should be excised, since these will be killed by the sterilant and may have adverse effects on callus development which occurs as a response to wounding (see below). The surface of clean sterile tissue can be lightly scored with a sharp blade to enhance cell division. If stems are being used, the ends can be dipped into beeswax prior to sterilization to prevent sterilant uptake. If necessary the explants can be dissected at this stage (e.g. for endosperm, radicals, embryos, etc.) and then transferred to the required sterile medium and incubated as necessary. If semi-solid media are used the explants should be gently pressed into the agar so that good surface contact is attained. The orientation of explant can be important; e.g. radical tips will form callus best if laid across the agar surface. Different orientations should be investigated.

Culture vessels should be clearly labelled with the date and information concerning the tissue. Labelling the container itself, rather than the lid, is useful if, by mischance, vessel lids become muddled during handling! Petri dishes may be sealed with 'Parafilm' to prevent desiccation during long incubation periods. Cultures for callus initiation are usually incubated at 25 °C in the dark. If initiation does not occur within 3–4 weeks, explants should be transferred on to fresh medium.

It should be noted that in plant cell culture the tissues are only *surface*-sterilized and the methods described above do not ensure against intracellular contamination by mycoplasmas, L-form bacteria, and viruses.

CHOICE OF CULTURE MEDIA

After the explant has been chosen and sterilized it has to be placed on a callus-inducing medium. Plant cell culture media are based on mixtures of at least 20

Table 1.1 Inorganic and organic compounds of four major plant culture media (mg L^{-1})

	Murashige and Skoog	Gamborg's B5 medium	White's	Heller's
Inorganic				
KH_2PO_4	170	—	—	—
KNO_3	1900	2527.5	80	—
KCl	—	—	65	750
Na_2SO_4	—	—	200	—
$NaNO_3$	—	—	—	600
$NaH_2PO_2.H_2O$	—	150	19	125
$Na_2MoO_4.2H_2O$	0.25	0.25	—	—
$Na_2EDTA.2H_2O$	37.3	37.3	—	—
$CaCl_2.2H_2O$	440	150	—	75
$Ca(NO_3)_2.4H_2O$	—	—	300	—
$MgSO_4.7H_2O$	370	246.5	750	250
$AlCl_3$	—	—	—	0.03
$Zn\ SO_4.7H_2O$	8.60	2	3	1
KI	0.83	0.75	0.75	0.01
$NiCl_2.6H_2O$	—	—	0.03	—
$FeCl_2.6H_2O$	—	—	—	1
$Fe_2(SO_4)_3$	—	—	2.5	—
$FeSO_4.7H_2O$	27.8	27.8	—	—
NH_4NO_3	1650	—	—	—
$(NH_4)_2SO_4$	—	134	—	—
$CuSO_4.5H_2O$	0.025	0.025	0.001	0.003
H_3BO_3	6.20	3	1.5	1.0
MoO_3	—	—	0.001	—
$MnSO_4.4H_2O$	22.30	—	5	0.1
$CoCl_2.6H_2O$	0.025	0.025	—	—
$MnSO_4.H_2O$	—	10	—	—
Organic				
Myo-inositol	100	100	—	—
Pyridoxine-HCl	0.5	1	0.01	—
Thiamine-HCl	0.1	10	0.01	—
Nicotinic acid	0.5	1	0.05	—
Glycine	2	—	3	—
Sucrose	3%	2%	2%	—

components which combine inorganic and organic elements (i.e. inorganic nutrients, trace elements, and iron source) together with a carbon substrate and appropriate plant growth regulators. Table 1.1 lists four of the most common media, although there are many others. Different plants require different nutrients, and it is known that callus derived from different parts of a plant may have different nutritional requirements. Indeed, although most callus and suspension cultures from the same explant can normally be grown on the same medium this is not always the case. Therefore, when a new system is being studied, a suitable medium for initiation and maintenance of both callus and suspensions can only be obtained by trial and error. Fortunately, many plants have been examined in tissue culture and usually some relevant literature is available. Dixon (1985) has provided details of media requirements for many plants commonly used for tissue culture, and precise details of setting up callus and suspension cultures. Another recent innovation which can simplify routine experimentation is that some plant culture media which require only the addition of a carbon source and plant growth regulators can now be obtained commercially. These mixtures are relatively expensive but are very useful when only a small quantity of a particular medium is required, and they save much time (and frustration) in media preparation. The same medium is usually used for both semi-solid and cell suspension cultures, with the gelling agent agar being added to semi-solid cultures. Commercially available agars are of variable purity and it is best to use relatively pure preparations (e.g. Oxoid Agar No. 1) at a concentration of 0.75% (w/v). The gelling strength of agars varies slightly between batches and with the ionic strength of the medium, and this will affect callus morphology. The impurities within agar, along with its very slight breakdown to galactose etc. during autoclaving, makes semi-solid culture inadequate for studying precise growth requirements. In this situation silica gel can be used as a setting agent, or *stationary liquid* cultures can be employed as an easier and better alternative.

The majority of media are now defined, so the quality and quantity of the medium components are known. However, non-defined components such as coconut milk and tomato juice are sometimes necessary in order to obtain good growth or when initiating callus. If possible, defined media should be used for subculturing so that consistency can be maintained. Non-defined plant extracts are prepared by homogenizing the material, extracting the juice, and filtering. These non-defined supplements should be prepared in large quantities, because this allows some uniformity for batch preparation. Coconut milk should be filtered through a coarse filter (e.g. muslin), boiled for 15–20 min to precipitate proteins, and then filtered through Whatman No. 1 paper. The milk should then be dispensed into suitable volumes and stored in a frozen state prior to use. Once thawed it can be added (usually at 10%, v/v) to the growth medium prior to autoclaving. The reasons for the improved growth characteristics associated with non-defined components are not totally understood, but are associated with the additional nutrients and/or plant growth regulators they contain. Coconut milk is also known to have some buffering capacity and so may protect the developing callus from pH fluctuations. Very little attention has been given to control of pH in plant cell culture, with the pH usually being adjusted to between 5.6 and 6.0 before autoclaving.

Plant growth regulators are of major importance, and the fine balance between auxin and cytokinin concentration may be crucial in establishing cultures and maintaining callus. Growth, differentiation, and metabolism have all been shown to be affected by plant growth regulators, and after initiating callus a study of their effects (on growth and secondary metabolite synthesis in particular) may be initiated. The results of one such investigation in which *Pelargonium* callus cultures

growing on Murashige and Skoog medium were exposed to varying concentrations of the synthetic analogue of kinetin (6-benzylaminopurine) and the
auxin naphthalene acetic acid (NAA) provide a good example of growth regulator
effects (Brown and Charlwood, 1986). Shoot growth from callus was enhanced on
relatively low (0.05–0.5 mg/L) concentrations of NAA, while root growth was
stimulated on media containing higher NAA concentrations (<5 mg/L). Preferential callus growth was observed on media containing high 6-benzyl-
aminopurine and NAA. In this case, as is a frequent observation with many plants,
the process of differentiation was greatly influenced by the concentration of
growth regulator present. However, some callus and cell suspensions become
habituated, i.e. the cells no longer require the addition of a particular plant growth
regulator for their maintenance and growth. Plant growth regulators can be kept
as concentrated stock solutions at 0–4 °C. Auxins are generally soluble in sodium
hydroxide, while cytokinins are soluble in aqueous ethanol. Since some plant
growth regulators are thermolabile, e.g. zeatin and gibberellic acid, they should
not be autoclaved but can be filter-sterilized and added to the sterile medium.

Sucrose (at 2–5% w/v) is the most common carbohydrate source in plant cell
culture media, although other sugars are also used. The carbon source may affect
both differentiation and secondary metabolism, as well as callus growth and
morphology. The use of other media components may be required for some tissues.
Thus, the adsorbents activated charcoal and polyvinylpyrrolidone can greatly
assist callus initiation in cultures which produce high levels of phenols.

CALLUS INITIATION AND SUBCULTURE

Callus initiation may be observed as an initial swelling at the areas of tissue
damage, with callus formation being typically observed within one to five weeks of
explant transfer on to a suitable medium. Since plant cells have a minimum
inoculation density, below which cells may remain viable but do not grow, the
callus should remain in contact with the explant until a sufficient amount (e.g.
minimum volume approx. $0.3 \, cm^3$) has formed. Subculture is without doubt
selective (both deliberate and unconscious) and selection can be considered at the
first transfer; e.g. the transfer of the reddest cells in selection for cells containing the
pigment betalain in beetroot. It is prudent to initiate a number of different cell
lines during the initial subcultures, when the degree of variation is probably at its
greatest. Subculture can be made using a sterile scalpel and forceps, although
friable callus does not usually require physical cutting and a spatula is often more
convenient.

Once rapidly growing cells are obtained, random selection of callus material for
subculture will allow for representative sampling. Cutting the callus into equal
proportions (cf. cutting a cake) will achieve some randomization, although it
should be remembered that cells in the centre of an older callus do not divide and
may be dead. The amounts of tissue subcultured should be kept relatively
constant. Different tissues do have different minimum inoculation potentials, but
as a guideline a $2 \, cm^3$ callus can be divided into four to six pieces. Typical
subculture periods range from three to five weeks, depending upon the culture

vessel and the tissue growth rates. Some care should be taken to subculture at regular intervals so as to maintain the stability of the cell line.

INITIATION AND SUBCULTURE OF CELL SUSPENSIONS

Initiation of callus and suspension cultures is a simple procedure for some species. From the microbiologist's point of view, a major drawback in plant cell culture is that plant cells have a minimum inoculation density. Thus a minimum number of cells (or the products of their metabolism) must be present in order for cell division to occur. If this minimum cell number is not achieved, a lag phase will be induced during which time the metabolites required for cell division will be synthesized. This characteristic is most significant when using cell suspension cultures, and subculture regimes should use cell numbers greater than that of the minimum in inoculum density. If large volumes of cells are required (see Chapter 9), scale-up regimes should be considered so that long lag phases are not induced. These scale-up regimes, along with the slow growth rate of plant cells may mean that in order to grow, say, 10 L of cells, the inoculum will have to be grown in a series of batches (say 0.1 L and 1.0 L) which will probably take four to five weeks. The nature of these cytokinetic metabolites produced during lag phase is not known, although single plant cells can be grown by supplementing fresh medium with medium which has supported growth of other plant cells i.e. a conditioned medium. This feature causes methodological difficulties in situations where single cells (often obtained from protoplasts) have to be grown and will be discussed in more detail in Chapters 3, 4 and 5.

Suspension cultures should be initiated using callus which has been maintained on semi-solid medium for at least six to ten passages, by which time the callus should be growing at a fairly consistent rate, indicating an adaptation to the medium. Dispersion is achieved by placing small sections of callus into liquid medium and shaking the culture on an orbital shaker. Friable callus is the easiest to initiate into liquid culture, as it will quickly disperse throughout the shaking medium. It is therefore advantageous to test different media in order to obtain friable callus prior to initiation of suspension cultures. Hard callus can be cut prior to addition into liquid, but will still generally take longer to initiate suspensions.

In cases where the suspensions become highly clumped, flasks with indentations which physically break up the aggregates can be used. Suspension cultures should ideally consist of single cells, but this is rarely the case and usually small aggregates 100–1000 µm in diameter are found. Large clumps of plant cells can be gradually eliminated from liquid cultures by filtering through a mesh of large porosity (e.g. 0.1–0.5 mm) prior to subculture; this is a particularly useful technique for the early stages of initiation of liquid cultures.

Suspension cultures typically grow faster than callus cultures, are morphologically more homogeneous, and hence offer certain advantages over callus cultures. Different laboratories use different protocols for subculturing plant cell suspensions, depending upon their research aims. At one level, subculture is utilized only to maintain a culture; at another it is used to maintain a stable cell line. Stability of cultures is obviously of great importance for reproducible experimentation, and is a characteristic which will be frequently discussed throughout this book.

It is best to subculture cells at a defined stage of growth, and this is usually performed towards the end of the exponential growth phase. Subculture is achieved by transferring a fixed volume of culture into fresh growth medium at constant time intervals in quantities greater than their minimum inoculation density, e.g. a volume ratio of 1:5 for *Catharanthus roseus*. For cultures that are clumped (e.g. aggregates >1 mm in diameter) aseptic transfer of filtered and weighed cells (e.g. 5 g wet weight, 100 mL) is more accurate. Again, in order to minimize variation it is advised that subculture takes place at regular intervals (e.g. 14 days). Since the maintenance of a particular cell line is of great importance it is strongly advised that some replicate flasks are always kept on a different shaker to protect against mechanical breakdown. As in callus subculture it is also worthwhile to take a note of the date and method of subculture along with any supplementary information that may be available concerning a particular cell line, e.g. growth rate, maximum biomass, product formation. This will allow any changes in the culture to be monitored so that culture procedures can if necessary be modified in addition to evaluating stability.

Plant cells are large, have a rigid cell wall, and usually become highly vacuolated during the later stages of the culture cycle. These features all combine to form a cell which is sensitive to hydrodynamic stress. Thus when cell suspension cultures are agitated, slow shaker speeds (usually 50–200 r.p.m.) have to be used so that the cells are not damaged. Such slow shaker speeds can result in poor mixing of liquid cultures and hence may greatly affect cell metabolism. These features of slow growth rate, minimum inoculation potential, and sheer sensitivity should be considered when undertaking any studies on plant cell and tissue culture.

Characteristics of Callus and Suspension Cultures

CALLUS GROWTH AND DEVELOPMENT

Callus provides a morphologically more uniform material than the whole plant, and is therefore a potentially useful model system for studying many areas of cell biology. It provides a material from which cell suspensions can be obtained, and recently initiated callus is often used as a source of regenerated plants. Over and above this, calluses are slow growing, small, and convenient to handle, and hence are a useful means of maintaining and storing germplasm. These features denote the importance of callus in both academic and commercially orientated research. Despite this there is surprisingly little information concerning callus development, with much of the research being conducted by Street and his colleagues (see Street, 1977). These researchers have recognized that callus development occurs in two phases; namely the so-called wound and growth responses. The *wound response* is characterized by limited cell division and a rapid increase in metabolic activity, but does not necessarily lead to callus development. The first report of a wound response was by Duhamel (in 1756) who observed the development of a swelling on a decorticized elm tree. The *growth response*, on the other hand, results in continued cell division and is usually dependent on an exogenous supply of auxin.

Thus, callus development is a natural response to injury with cells around the area of damage dividing rapidly so that a layer of cells forms over the wound.

Increased metabolic activity is involved in such processes (e.g. polyphenol production to strengthen cell walls) and compounds are synthesized which protect against pathogens, e.g. the enzyme chitinase which hydrolyses the fungal cell wall component chitin. The reason why formerly quiescent (i.e. differentiated and non-dividing) cells are induced to divide and form this protective layer of cells (i.e. wound cambium) is not known, but it has been postulated to result from the interaction of hormone(s) liberated by the damaged cells. Practically speaking, this information should be considered when excising explants, since one with a high ratio of surface area to volume will be most likely to make effective hormone available. Indeed, many researchers advocate that the explant should be lightly scored with a scalpel in order to encourage the wound response.

The callus growth response has been divided into three stages which are characterized by changes in mean cell size of the population and overall metabolism, and culminates in the appearance of a callus. These stages have been termed *induction* (when cells 'prepare' for division as seen by metabolic changes and the occurrence of macromolecular synthesis), *division* (when cells become meristematic, actively divide and consequently their mean cell size decreases), and, somewhat controversially, *differentiation*. In this stage the cells begin to enlarge and become vacuolated, and the rate of cell division also decreases so that an equilibrium is reached between cell expansion and division. Some researchers strongly object to the term differentiation being used for this stage because it refers only to cytoplasmic changes. Differentiation is more often used to describe changes in cell morphology which lead to the formation of a cell with an associated specialized function, e.g. xylem and phloem.

Both the wound and growth responses are involved in the transformation of specialized cells into meristematic cells which ultimately results in callus formation. This is often evident in callus initiation at excised (i.e. wounded) areas of tissues, e.g. the ends of stems, around the edges of a leaf disc.

Caplin (1947) made an investigation of callus growth by examining the change in external morphology of precisely measured cultures of tobacco callus. Figure 1.2 shows some of his results, illustrating that most rapid growth occurs at the points where there is contact between the cells and growth medium. Moreover, as growth progresses its rate varies in different areas within the callus. The proliferation of material at the periphery of the growing callus has been related to the wound response, although there appears to be some similarity between callus growth and that of the microbial colony. In bacteria and fungi, colony extension is, after an initial logarithmic phase, restricted to an outer peripheral growth zone at a rate which is dependent on the rate of supply of the growth-limiting nutrient; thus the radial growth rate of colonies is linear. Caplin's results show similar patterns, although more detailed research and analysis is required. Another feature of callus growth is that the increase in wet (fresh) weight and cell number usually occur at an exponential rate so that growth is balanced.

HETEROGENEITY OF CALLUS

The key role of callus in plant cell culture has already been stressed, as has the importance of maintaining callus in controlled conditions. However, irrespective

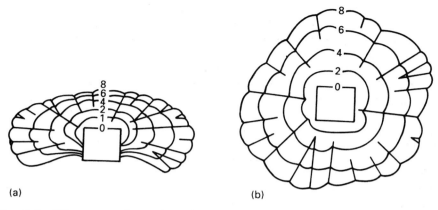

(a)   (b)

Fig. 1.2 Diagrammatic sections through a callus culture of *Nicotiana* sp. illustrating morphological changes during 8 weeks' growth. A, section perpendicular to nutrient agar surface; B, section parallel to nutrient agar surfaces. Cubes changed into somewhat flattened hemispheres by growth of knobs in which proliferation of cells occurred in and near surface. Concentric lines represent shape of culture after indicated number of weeks' growth. Radial lines indicate surface of contour between adjacent knobs. Points of radial lines nearest centre are points at which knobs have divided. Where two lines radiate from such points, new knobs have been formed. (From Caplin, 1947, with permission.)

of the control imposed, there is heterogeneity both between calluses and within a callus itself. This heterogeneity is seen in established calluses as differences in colour, morphology, structure, growth, and metabolism. Even an apparently uniform callus (in terms of visual appearance) may contain cells of different ploidy and metabolic capability. Callus heterogeneity can arise from:

- Different genotypes within a species.
- The different tissue (and cell types) from which callus is initiated.
- The conditions and methods in which calli are maintained and subcultured.

Callus from mixed cell types (e.g. root tissue) will produce a mixed callus. In many cases one particular cell type will have a faster growth rate and will eventually dominate the callus. In some cases this is controllable. In one example pea root was cultured on a medium containing the auxin 2,4-D and yeast extract; the resultant callus contained cells which were diploid, tetraploid, and octoploid. When the yeast extract was omitted the callus consisted only of diploid cells (Burgess, 1985).

The morphology of callus is dependent on its mode of growth. As callus grows, cells are pushed upwards and outwards from the surface of the medium so that nutrient gradients are established between the cells and the growth substrate. These gradients will involve nutrients (including gases), plant growth regulators, the products of metabolism, etc., and presumably micro-environments are established. Such gradients and micro-environments will in themselves affect growth patterns, e.g. by promoting and inhibiting cell expansion and division. It is impossible to prevent this variation in any closed system where nutrients are not being replenished nor metabolites removed. This heterogeneity within a single callus illustrates the importance for random selection during subculture (see page 11), while heterogeneity between calluses can be reduced by using strict subculture protocol. However, two calluses derived from the same explant but cultured on two different media should be considered as two different cell lines. This heterogeneity implies that no two calluses are identical. However, calluses do become less variable with time and so some consistency can be attained.

HETEROGENEITY OF SUSPENSION CULTURES

The metabolic heterogeneity of callus cultures as described above is greatly reduced in suspension cultures because all of the plant cell is submerged in the nutrient medium. Unfortunately true homogeneity is not achieved since, in liquid culture, plant cells have a tendency to grow as aggregates. Single-cell suspensions have been reported, as have some methods for their attainment. However, in most cases these are not long-lasting and small aggregates eventually form. The quality of a suspension culture is therefore described as 'fine' or 'aggregated', depending upon the aggregate size. However, such definitions are arbitrary and in research directed towards secondary metabolite synthesis, for example, there is some debate as to the quality of suspension culture which should ultimately be obtained. While, in theory, finely suspended cells should be totipotent and capable of synthesizing the broad range of metabolites produced by the whole plant, in practice accumulation of such compounds does not always occur. On the other hand, large clumps of cells often contain cells which are to some extent differentiated, a state which according to some researchers is a prerequisite for secondary metabolism. Arguments for each line of reasoning are available, and there appears to be no definitive answer. Large clumps of cells are, in liquid culture, probably analogous to callus material in that heterogeneity is enhanced by nutrients becoming limited towards the centre of the aggregate. It would seem reasonable to speculate that this depletion of nutrients would have dramatic effects on metabolism.

GROWTH OF PLANT CELL SUSPENSIONS

Plant cells will, in suitable growth conditions, increase their size and ultimately divide with the formation of two cells capable of further growth. Thus cell growth consists of expansion and multiplication. Situations frequently arise where cell expansion occurs but is not followed by cell division. Indeed, plant cells frequently maintain viability but do not exhibit any features of growth. This can be due to a variety of reasons e.g. depletion of an essential nutrient, the presence of toxic materials, transition from one physical environment to another (drop in pH), etc.

  Growth can be considered at the level of the individual cell (i.e. the cell cycle) or the population (i.e. growth cycle) and it is the techniques involved in measuring the latter which are important for routine analysis of cell suspension cultures.

Measurement of Growth of Cultured Plant Cells

Growth can be followed using the parameters applied to microbial cultures. However, the success of a particular technique will depend on the nature of the plant cell suspension, e.g. aggregated versus fine cell suspensions. The following techniques can be used to determine growth of plant cell cultures:

- wet (fresh) and dry weight
- cell number
- cell viability
- packed cell volume
- cellular protein and DNA content
- mitotic index
- medium components, e.g. carbohydrate, nitrate, phosphate
- medium conductivity

WET (FRESH) AND DRY WEIGHTS

Determination of plant cell biomass, where the weight of living material is estimated, is the most common parameter for growth measurements. This can be measured directly by making fresh weight estimations; indirect methods, which are often non-destructive, offer advantages especially in mass cultivation systems.

Wet (fresh) and dry weights are determined by using a filter unit consisting of a stainless steel mesh placed in a Buchner funnel to which gentle suction can be applied. Filter papers can be supported on this mesh, and if required a metal ring can be made to facilitate handling of deeper beds of filtered cells. Dry weights of the filter papers are taken after drying the filters in an oven at 60°C until a constant weight is achieved (approx 18 h). Wet weights are then determined by placing the dried filter paper on the filter unit, wetting it, removing the excess water, and reweighing. Known volumes of evenly dispersed, aseptically sampled, cell suspension are then filtered. The cells are washed with distilled water and then re-weighed together with the filter paper to determine wet weight (expressed in mg cells/mL of suspension culture). Dry weight measurements are made by drying the cells plus filter until a constant weight is attained (18 h). These measurements are usually expressed as g/L, but this may be misleading since growth characteristics may be dependent on culture volume. Three measurements for each estimate (expressed as mean ± standard error) are necessary for a representative determination. Sampling at three-day intervals is usually sufficient for obtaining a growth curve in a typical plant cell suspension, although more frequent sampling may be necessary during the exponential growth phase. Media samples can be saved (stored at below 0°C) for analysing changes in the concentrations of various medium components during growth. Wet weights are difficult to assess, and care should be taken to maintain a consistent methodology (e.g. suction pressure, duration of applied suction). Dry weights may be misleading due to accumulation of storage products such as starch and formation of extracellular polysaccharides, and a cross-check with cell number determinations should be made before routine analysis is based on this technique.

The mathematical expression of growth rate is important when dealing with a changing cell population as observed in batch cultures (i.e. a culture in a finite volume of medium in a system which is closed except for gaseous exchange). As with other cells, plant cells in batch culture go through lag, exponential, stationary, and death phases. These phases obviously reflect the continually changing environment where growth ultimately depletes the nutrients in the culture medium and metabolites are produced.

The rate of plant cell growth is, under certain conditions, proportional to the cell concentration already present (i.e. it is autocatalytic). Under ideal conditions the cell biomass increases in a logarithmic (exponential) manner, i.e.

$$N_0 \to 2N_0 \to 4N_0 \to 8N_0 \to nN_0$$

where N_0 is the wet weight, dry weight, or cell number, at $t = 0$. In order to facilitate explanation, let N be the cell number. Under ideal conditions, when exponential growth is occurring, the rate of increase in cell number is constant as is the interval taken for the cell population to double. This latter interval is termed the doubling time (t_d) and in this situation is equivalent to the mean generation time (MGT). The MGT is the average period for one complete cell division cycle.

The number of doublings n after time t is given by

$$n = \frac{t}{t_d}$$

It follows that the cell number after time t will be related to the number of cells originally present (N_0) and therefore

$$N_t = N_0 2^n = N_0 2^{t/t_d}$$

Rearranging the above equation to incorporate natural logarithms (i.e. $\log_e$),

$$\ln \frac{N_t}{N_0} = \frac{\ln_2 t}{t_d}$$

Thus, since the natural logarithm of 2 is 0.693,

$$\frac{\ln N_t - \ln N_0}{t} = \frac{\ln 2}{t_d} = \frac{0.693}{t_d} \tag{1}$$

A graph of $\log N$ against time will, in exponential growth, give a straight line. The gradient of this line is described by equation (1) and is called the specific growth rate (μ). During exponential growth, $\mu = 0.693/t_d$.

Specific growth rate is usually calculated from plots of $\log N$ versus time, t, with linear regression analysis being used to determine the 'line of best fit'. Figure 1.3(a) and (b) shows growth curves of a cell suspension of *Picrasma quassioides* plotted on arithmetic and logarithmic scales. In the lag phase there is no increase in cell number, and as discussed above, the duration of this phase is dependent on the initial inoculum concentration. After a period, the cells will begin to divide and after an acceleration phase the culture will enter into exponential growth. Subsequently, and usually as a result of a nutrient becoming limiting, the rate of

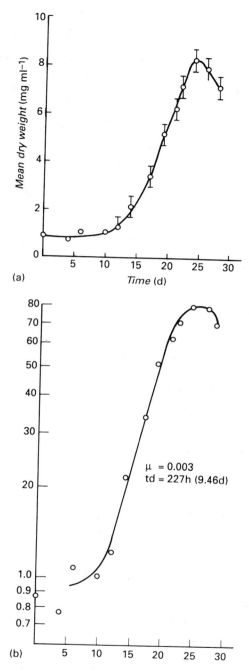

Fig. 1.3 Growth of *Picrasma quassioides* cell suspension culture. The same data is shown plotted on arithmetic (a) and semi-logarithmic (b) scales. Note the extremely slow growth rate and increasing standard errors caused by the culture becoming more aggregated.

cell division will decrease so that the culture will enter into stationary phase. Finally, and as a result of nutrient depletion or accumulation of growth inhibitor, the cells will, in batch culture, begin to die.

When assessing the performance of plant cell suspensions for the production of for example secondary products or enzymes, it is of vital importance to obtain estimations both of maximum biomass achieved and of specific growth rate. Maximum biomass expressed in terms of dry cell weight/volume culture can be obtained from growth data plotted arithmetically or on a semi-logarithmic scale, so long as frequent samples are made before and after the onset of stationary phase. With respect to estimating the specific growth rate of plant cell suspensions, one problem which frequently arises is that, generally because of aggregation, the exponential phase is difficult to determine. Some researchers therefore calculate the specific growth rate between the points (sometimes only two or three) where the fastest increase has occurred. This results in much faster growth rates, but is an unrealistic measurement. Instead, it is possible to calculate the growth rate between the range of points where N (e.g. dry weight) begins to increase and subsequently begins to decrease. Again, it is essential to sample the culture frequently throughout the growth phase (preferably every 24 h).

An additional method for determining biomass concentration is via the measurement of turbidity using either spectrophotometry or nephelometry, although the aggregated nature of plant cell suspensions makes this technique inaccurate in the present context. Indirect methods such as oxygen balancing may also be used, and these offer advantages in non-homologous systems such as root cultures. Cell number and packed cell volume measurements are also commonly used to estimate biomass concentrations, and these methods are further described below.

CELL NUMBER

Growth can be determined in terms of a number of cells per litre. Cell numbers of very fine cell suspensions can be determined directly with the aid of a microscope using a Neubauer haemocytometer. (A Thomas counting chamber is generally too small for plant cells, with a chamber depth of only 0.02 mm.) However, such cell suspensions are rare and it is usually necessary to break down the cell aggregates so that individual cells can be observed. Samples can then be measured using a counting chamber although direct counts from known volumes are more straightforward.

Cell aggregates can be broken down by adding 1 volume of evenly dispersed cell suspension to 2 volumes of 8% (v/v) chromium trioxide solution in a siliconized Universal bottle. Samples can be stored in a refrigerator for one to six weeks (depending upon the particular cell suspension) or handled more quickly by heating to 70°C for two to fifteen minutes (again depending on the particular cell suspension) and allowing to cool. Samples are mixed and diluted with known volumes of water using accurate pipettes. (The ends of plastic pipette tips should be excised to ensure accurate sampling.) The suspension is drawn through a plastic syringe with a needle (No. 16 B-D) to ensure cell separation. Droplets of known volume, e.g. 10 μl, of the well-mixed suspension can then be counted using low-power light microscopy. Dilutions should be made so that there are approximately 200 cells per drop (10 µl) with at least five drops being counted for each sample (analysis of the error obtained in making counts will determine the number of cell counts which have to be made). Replicate dilutions should also be made for each stored sample. Since plant cells adhere to glass, all dilutions should be made in siliconized glass or plastic containers. Alternatively, cell aggregates may be broken down using 0.2% (w/v) pectinase in a buffer with some osmotic support rather than by use of chromium trioxide. The advantage of chromium trioxide is that it can be stored ready to use and is much cheaper. Storage of treated samples until successful completion of the experiment may also be advantageous (especially in cases where cultures become contaminated late in the growth cycle) so that needless tedious cell counts can be avoided.

Automated methods for cell number determination are available (i.e. Coulter counters) but these have obvious limitations, over and above their expense, when working with typical cultured plant cells which do not separate after division.

CELL VIABILITY

The presence of a semi-permeable cytoplasmic membrane allows viable plant cells to be identified by their ability to either accumulate or prevent the uptake of certain stains. Thus, enzymatic hydrolysis of the fluorogenic substrate fluorescein diacetate results in the accumulation of fluorescein, while treatment with Evan's blue results in staining of non-viable cells. Fluorescein diacetate can be prepared as a stock solution (5 mg/mL in acetone) and stored at 0–4°C for several months. Working aqueous solutions (0.1 µg/mL) are used to directly stain cells which are observed using fluorescent microscopy. Viable cells exhibit a bright green fluorescence under ultraviolet light (360 mm) after approximately 2 minutes staining, while non-viable cells do not fluoresce. Viability can be expressed as a percentage of the total cell population observed, and one can very quickly become experienced at estimating these percentages to determine the quality of a suspension culture.

Viability can also be qualitatively assessed by microscopic examination of intact cell membranes, nuclei, and cytoplasmic streaming. However, a much more accurate measurement can be made using a method described by researchers at the Institute of Food Research, Norwich (Parr *et al.*, 1984). The total intracellular volume is determined by mixing a freely permeable substrate (tritiated water) and an impermeable substance (^{14}C-mannitol) with a sample of cell culture. The two substances are diluted to different degrees depending upon membrane integrity (and hence cell viability) and the total intracellular volume of intact (viable) cells can be calculated. This parameter can be used as a measurement of cell growth.

Viability is an important characteristic but it should be remembered that it is not a parameter of growth (i.e. cell division and expansion), only of survival.

PACKED CELL VOLUME (PCV)

This method measures the volume of cells present at any time during the growth cycle. It is a much less time-consuming method than those described above, and provided that the suspension culture is relatively fine and the methodology consistent, then it is also very accurate. Once the PCV measurement is achieved, the same cells can be used to determine wet/dry weight measurements or cell numbers.

Small volumes (up to 10 mL) of suspension culture are aseptically sampled and placed in a graduated conical centrifuge tube. The total volume of a cell pellet is determined after centrifuging at a constant speed (e.g. $500\,g$) for a specific time (e.g. 5 min). The PCV is typically expressed as a percentage of the total volume in the tube.

CELLULAR PROTEIN AND DNA

The amount of a cell component, which is a constant proportion of the total cell dry weight, can be used to estimate biomass concentration during balanced growth. Components such as protein and DNA are usually determined by using well established techniques e.g. by Bradford's reagent for protein. Cell nitrogen, RNA, and ATP are now, with the advent of automated methods, also becoming popular parameters of culture growth.

MITOTIC INDEX

Cell division is obviously an integral part of growth. In eukaryotes, there are two types of cell division: *mitosis* (the normal cell division process leading to growth), and *meiosis* (permitting sexual hybridization while retaining the same DNA complement as the parent cell). Mitosis is characterized by a number of stages termed prophase, metaphase, anaphase, and telophase. Finally, cell division occurs. Mitotic index is an estimate of the number of cells of the population in these stages (i.e. mitosis). Determination of mitotic index is simple, but, like cell number estimations, it is time consuming. However, used in conjunction with cell number estimation it is a valuable parameter.

The mitotic phase is easily distinguished from the interphase where the cell chromatin is diffuse, because it is only during mitosis that individual chromosomes can be seen under light microscope. Observation of chromosomes is usually facilitated by the use of one of a wide range of DNA stains which are readily available.

The method described uses the fluorescent dye diamidino-2-phenylindole (DAPI) which specifically binds to DNA, although other dyes (e.g. Feulgen, Toluidine Blue) may also be used. Cells are washed twice in distilled water and the pellet fixed for 1 h at room temperature in 2 volumes absolute ethanol:1 volume of glacial acetic acid. Fixative is removed by centrifugation and the cells washed twice in distilled water. Cells are then stained using $0.5\,\mu g/mL$ DAPI (a stock solution of $50\,\mu g/mL$ can be stored in the dark at $0-4°C$ for several months) and mounted on a microscope slide with a coverslip. The slide is then placed between two pieces of filter paper and tapped vigorously (but carefully) and so squashed. Observation using fluorescence microscopy (exciter/barrier filters =

400 nm) shows nuclei and chromosomes staining bright blue against a black background.

Approximately 1000 nuclei should be scored, counting those that are obviously in some phase of mitosis.

$$\text{Mitotic index} = \frac{\text{No. of nuclei in mitosis}}{\text{Total no. of nuclei scored}} \times 100$$

If cells are highly aggregated they can be broken down after fixing, by using a 2 h digestion in 1% pectinase (in 0.1 M sodium acetate buffer, pH 4.5).

Mitotic index is directly proportional to the duration of mitosis/duration of average cell cycle, provided all the cells in the culture are going through the cell cycle. In plant cell cultures the duration of mitosis is usually short relative to the total cycle time (< 10%). Thus, a peak in mitotic index indicates an initial, rapid shortening of the mean cell cycle (i.e. mean generation time).

MEDIUM COMPONENT CALIBRATION

The depletion of medium components, e.g. glucose, nitrate, ammonium, phosphate, can be used to estimate cell growth. Media fractions can be collected during wet/dry weight analysis, frozen and assayed as required. Care should be taken in analysing results (especially for ions such as phosphate), as plant cells may possess large intracellular pools.

CONDUCTIVITY OF MEDIUM

The integrity of the plant cell membrane is reflected in the concentration of solutes which leak into the medium. Thus, the electrical conductivity of the medium can be used to analyse cell growth. There is an inverse relationship between medium conductivity and growth (i.e. higher conductivity during the lag phase compared to the exponential). Conductivity measurements have been used to determine seed vigour (Powell, 1986), but such simple and rapid methods have surprisingly not been applied generally to plant cells in culture. These measurements may be particularly useful when considering experiments on shear sensitivity and so could be very useful for studies on mass cultivation.

References and Further Reading

Brown, J.T. and Charlwood, B.V. (1986). Differentiation and monoterpene biosynthesis in plant cell culture. In *Secondary Metabolism in Plant Cell Cultures* (Morris, P., Scragg, A.H., Stafford, A., and Fowler, M.W., eds) Cambridge University Press, Cambridge.

Burgess, J. (1985). *An Introduction to Plant Cell Development.* Cambridge University Press, Cambridge.

Caplin, S.M. (1947). Growth and morphology of tobacco tissue cultures *in vitro*. *Botanical Gazette* **198**, 379–393.

Dixon, R.A. (ed.) (1985). *Plant Cell Culture, A Practical Approach.* IRL Press, Oxford.

Gautheret, R.J. (1985). History of plant tissue and cell culture: a personal account. In *Cell Culture and Somatic Cell Genetics of Plants* (I. Vasil, ed.), Vol. 2, Academic Press, London, pp. 1–59.

Jones, L.H. (1983). The oil palm and its clonal propagation by tissue culture. *Biologist* **30**(4), 181–188.

Parr, A.J., Smith, J.I., Robins, R.J., and Rhodes, M.J.C. (1984). Apparent free space and cell volume estimation: a non-destructive method for assessing the growth and membrane integrity/viability of immobilised plant cells. *Plant Cell Reports* **3**, 161–164.

Pirt, S.J. (1975). *Principles of Microbe and Cell Cultivation*. Blackwell Scientific Publications, Oxford.

Powell, A.A. (1986). Cell membranes and seed leachate conductivity in relation to the quality of seed for sowing. *Journal of Seed Technology* **10**, 81–100.

Stanier, R.Y., Adelberg, E.A., and Ingraham, J.L. (1977). *General Microbiology*, 4th edn. Macmillan, New York.

Street, H.E. (1977). *Plant Cell and Tissue Culture*, 2nd edn. Blackwell Scientific Publications, Oxford.

Chapter 2

Genetics of Cultured Plant Cells

ANGELA STAFFORD

Genetic Characteristics of Plant Cells

The particular genetic characteristics of plant cell cultures can only be discussed in the context of what is known about the genetics of higher plants in general. It is therefore appropriate to start this chapter with a brief summary of current knowledge relating to the structure and organization of the plant genome. This will provide a sound basis for the remaining sections in which we deal with phenotypic and genetic variability in cultured plant cells, a phenomenon which has far-reaching implications for the use of plant cell culture in agriculture and industry. Some general texts covering the area are those by Leaver *et al.* (1986), Mantell *et al.* (1985), Nagl (1978), and Yeoman (1986).

THE NUCLEAR GENOME

Nuclear DNA Content Diversity and Organization
The amount of DNA per haploid nucleus varies widely between species; for example *Triticum aestivum* (wheat) haploid nuclei contain 15.7 pg DNA packaged into 21 chromosomes, whereas *Saccharum officinarum* (sugar-cane) haploid nuclei have only 4.3 pg DNA which is organized into 40 chromosomes, and in *Arabidopsis* 0.8 pg DNA is packaged into five $(2n = 10)$ chromosomes. Even within a family, variation in DNA level and chromosomal organization can be vast. In the *Leguminosae*, for instance, the broad bean (*Vicia faba*; $n = 6$) and the runner bean (*Phaseolus coccineus*; $n = 11$) have haploid nuclear DNA contents of 14.4 pg and 1.7 pg respectively.

Plant cells contain comparatively large amounts of DNA, even the smallest plant genome (*Arabidopsis*, at 70 000 k bp) being several times larger than that of the fruit fly, *Drosophila melanogaster*. It has been calculated that in every case, the genetic material carried within plant cells is far in excess of that required to code for all the proteins synthesized during the growth and differentiation of the whole organism. For instance, in mung bean only 4% of the total genome is estimated to be coding sequence. The true function of this 'extra' DNA is not understood; however, studies carried out over the past decade or so have made some progress towards an understanding of nuclear gene organization. One of the most popular investigative techniques to have been used by biochemists involves denaturation of sheared double-stranded DNA molecules into single strands. These are then subjected to renaturation conditions (e.g. 60° C in 0.18M NaCl), under which the complementary regions of the single strands are expected to reassociate. Investigations of the *renaturation kinetics* of the DNA extracted from many organisms has revealed that a large percentage of the genome reassociates much more quickly than if only single-gene copy sequences were present. For example, it has been estimated that in wheat, at least 75% of the genome consists of repeated sequences which have probably arisen via sequence amplification during the evolution of the species. The non-repeated DNA fraction contains not only diverged repeated DNA (i.e. with accumulated mutations) but also structural gene sequences. In cereals, structural genes are expected to occupy less than 1% of the genome. Of the remainder, the structural genes for ribosomal RNA have been extensively studied; these are highly reiterated and are located at chromosomal sites termed *nucleolar organizers*.

Analysis of some plant structural genes including those specifying ribosomal RNA has been made possible by the development of gene cloning procedures. Comparison of cDNA and genomic DNA sequences representing a range of plant proteins has revealed that, as with many higher organisms, some plant structural genes contain coding sequences interspersed with intervening sequences (*introns*). Introns are processed out during mRNA maturation and are not therefore translated. The ribosomal RNA genes do not contain introns, but several other structural genes do, for example the leghaemoglobin gene from soybean and the phenylalanine ammonia-lyase (PAL) gene of *Phaseolus vulgaris* (French bean).

The Cell Cycle

Chromosomes can be visualized in plant cells using a variety of methods, including phase-contrast and interference microscopy on living plant cells or protoplasts, or DNA-specific dyes such as Feulgen or fluorescent dyes such as DAPI (diamidino-2-phenylindole). During mitosis, the interaction of histone and non-histone proteins with nuclear DNA produces the conformational state visible as chromosomes.

By performing pulse-chase experiments with tritiated thymidine, it has been possible to calculate the time taken for cells to undergo one cycle of growth and division. Cell cycle times vary widely according to the position of the tissue in the whole plant, and also depend upon whether or not the tissue is cultured. In the few instances where direct comparison has been possible between cultured and *in vivo* cells, it is obvious that cultured cells generally have very extended cell cycle times,

Table 2.1 Cell cycle length variation in plants and cell cultures (from Bayliss, 1985)

Species	Cell cycle phase duration (h)				
	T	G_1	S	G_2	M
Daucus carota (root tip)	7.5	1.3	2.7	2.9	0.6
Daucus carota (cell culture)	51.2	39.6	3.0	6.2	2.4
Zea mays (root tip)	9.9	1.7	5.0	2.1	1.1
Zea mays (cell culture)	37.0	23.9	7.1	3.9	2.1

T, total cell cycle length; G_1, synthesis of enzymes necessary for DNA replication; S, bidirectional replication of DNA and histone synthesis; G_2, cell growth, synthesis of proteins required for mitosis; M, mitosis during which chromosomes are distributed equally to two daughter cells.

this being largely due to an increase in the length of the G_1 phase (Table 2.1). This difference between plant and culture to some extent reflects the very different cell morphology of *in vivo* meristematic cells (small, densely cytoplasmic) and typical cultured cells (larger, more vacuolated). The suggestion has been made that the levels of auxin required to suppress differentiation in cell cultures are inhibitory to the progression of the cell cycle; certainly when cultured tissues regenerate, the total cell cycle time is reduced.

Gene Expression

The elucidation of the sequence of events culminating in the expression of a polypeptide *in vivo* has relied mainly on the use of prokaryotic and animal rather than plant systems, and only in recent years has any understanding been reached of the mechanisms controlling transcription and translation in plants. It would be expected that the regulation of gene expression in plants would differ in some aspects at least from animal systems, for a number of reasons. For example, whereas major morphogenesis in animals occurs during embryo formation, in plants meristems capable of differentiating into new organisms are maintained throughout the life cycle. In addition, plants are highly adaptive to external stimuli such as light, water conditions, and temperature, and these factors all play a major part in the development of the organism.

It has been suggested that at least 100 000 different genes are expressed during the life of a plant. Comparative studies on mRNA and nuclear RNA profiles from different plant tissues and organs indicate that gene expression is highly regulated at the transcriptional and post-transcriptional level, that is at the level of RNA accumulation. Experiments in which the application of external stimuli bring about consistent developmental responses have yielded some particularly interesting results. For instance, the treatment of soybean seedlings with auxin has been shown to enhance the levels of only a few mRNAs out of a population of several thousand mRNA species. These increases in mRNA levels were most pronounced in the region of cell elongation, and could reflect either an enhanced transcriptional rate or decreased mRNA turnover. Of particular relevance to those with an interest in the expression of plant secondary metabolism in cell cultures is the microbial elicitation of phytoalexin production. For instance, in French bean suspension cultures, polysaccharides prepared from cell walls of a fungus (*Colletotrichum*) selectively induce a rapid and transient induction of enzymes synthesis, brought about by the accumulation of specific mRNAs (Cramer *et al.*, 1985). Those enzymes induced include a number with a key regulatory role in the biosynthesis of isoflavonoid compounds which are thought to have an antimicrobial activity (see Chapter 6). Although it is considered that the principle level of control is mRNA accumulation, post-translational controls as well as transcriptional controls are suspected to operate in this process.

The phenylalanine-ammonia lyase (PAL) gene has become a target for the study of gene organization and structure because it can be induced via rapid mRNA synthesis to high levels in the plant and in cell cultures by the application of various environmental triggers (see Chapter 6). Recent studies (Cramer *et al.*, 1989) have revealed that three classes of PAL genes were present in the *Phaseolus vulgaris* genome, within which a number of diverse forms could be detected. When the nucleotide sequences of two PAL genes were compared, it was found that one contained an open reading frame encoding a polypeptide of 712 amino acids, while the second encoded a polypeptide of 710 amino acids, showing 72% similarity with the first. Both genes contained introns, at the same location. Extensive sequence divergence was found in the intron and in the 5' and 3' flanking regions. The 5' flanking region of both genes contained 'TATA' and 'CAAT boxes', promoter sequences in common with other eukaryotic structural genes, as well as elements resembling viral (SV40) coding sequences. These latter transcription regulation sequences have also been found in the promoters of other plant genes including those encoding octopine synthase, alcohol dehydrogenase, proteinase inhibitors, and chalcone synthase. No clustering of PAL genes was detected in the bean genome, in contrast to the chalcone synthase family, in which six to eight genes are clustered, as well as being extensively conserved within both coding and intron regions.

The divergent structural nature of these PAL genes was reflected in a number of regulatory features. For instance, in cell suspension cultures it was found that not all of the PAL forms were activated at the gene level by fungal elicitor, and differences were also observed in their response to irradiation and infection in the whole plant. These studies of the PAL gene family are providing some fascinating insights into the environmental and developmental regulation of transcriptional activity in plants.

Much effort is currently going into the identification of control sequences which allow genes to respond to defined developmental situations. In this context, transgenic plants have been produced (for example, tobacco plants containing soybean lectin genes) in which it has been demonstrated that foreign plant genes can be expressed in the correct differentiated tissue and with the expected developmental sequence. This type of experiment demonstrates that regions adjacent to (often upstream from) genes are responsible for controlling gene expression and that these sequences have been conserved in different species.

THE CYTOPLASMIC GENES

While the nucleus contains the majority of plant genes, the mitochondria and chloroplasts also contain functional DNA; the day-to-day running of a plant cell in fact relies upon the interaction of these three genomes.

Mitochondria

Mitochondrial (mt) DNA comprises less than 1% of the total cell DNA, but carries structural genes for several essential polypeptides including three subunits of cytochrome c oxidase, two subunits of the ATPase complex, and seven subunits of the NADH-ubiquinone oxidoreductase complex (Fig. 2.1). The genomes of plant

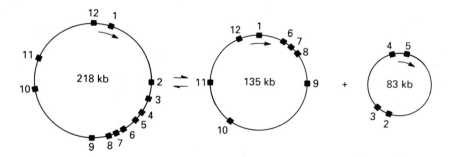

Fig. 2.1 Organization of the mitochondrial genome in *Brassica campestris* (as determined by hybridization to DNA probes prepared from other plant species) (from Makaroff and Palmer, 1987).

Gene		**Product**
1, 5	*cox II*	cytochrome oxidase subunit II
2	*atp 6*	subunit 6 of F_0 ATPase complex
3	*cox I*	cytochrome oxidase subunit I
4	*cob*	apocytochrome b of subunit II
6	*cox III*	cytochrome oxidase subunit III
7	*rps 13*	ribosomal protein S13
8	*ndh 1*	subunit 1 of NADH dehydrogenase complex I
9	*atp 9*	subunit 9 of F_0 ATPase complex
10	*atp A*	alpha subunit of F_1 ATPase complex
11	*26s rRNA*	rRNA
12	*18s + 5s rRNA*	rRNA

mitochondria are larger than those from fungi and animals (16 kb for mammalian mtDNA), and encode at least 20 polypeptides; animal mtDNA only carries in the region of 13 structural gene sequences. The phenomenon of cytoplasmic male sterility (CMS) which has been exploited by plant breeders for a number of years to produce hybrid crop plants, has been correlated with alterations in the mitochondrial genome.

mtDNA is highly variable in both size and form; in plants, it can be either circular or linear and can vary between species from around 200 kbp in brassicas to more than 2500 kbp in cucurbits. Restriction mapping has indicated that mtDNA consists mostly of unique sequences, and that these are carried on different-sized circles which are capable of recombination at specific sites of direct repeated sequences (see Fig. 2.1). It has been found that while the mitochondrial genomes from closely related plant species tend to be highly conserved in their primary DNA sequence, they do vary in linear gene order. *Brassica* species have relative small mitochondrial genomes and as a consequence these have been extensively studied; the mitochondrial genome of *Brassica campestris* consists of a 'master' circle of 218 kbp which recombines through a short direct repeated sequence to form two additional circles of 135 and 83 kbp. It has been suggested that the 'master' circle in this species is the only DNA molecule capable of replication, a hypothesis which would fit with the relative stability of plant mitochondrial genomes; even if recombination occurs frequently and randomly, any resulting subgenomic molecules would not be passed on to progeny mitochondria, only those containing the full complement of mt genes.

Sequences homologous to chloroplast DNA have been found in plant mito-chondrial genomes; for example, maize mtDNA has been observed to contain one sequence highly homologous to the ribulose-1-5-bisphosphate carboxylase large subunit gene, encoded by chloroplast DNA. From DNA sequencing studies it is apparent that the transfer of chloroplast DNA to mitochondria in higher plants is common; in addition, nuclear DNA has been found to possess sequence homology with both mtDNA and chloroplast DNA.

For further details of plant mitochondrial genomes see a recent review by Newton (1988).

Chloroplasts
The presence of DNA and ribosomes in chloroplasts was demonstrated in 1962, and it has since been shown that these and other plastids contain all the machinery required for gene expression. A typical leaf cell may contain 20–50 chloroplasts, contributing as much as 15% of the total cellular DNA content. In contrast to mtDNA, chloroplast DNA (ctDNA) appears to be relatively conserved between species, consisting of a multicopy circular molecule of approximately 150 kbp. In all, it is estimated that the chloroplast genome occurs in multiple copies of 19000–50 000 per cell. With a few exceptions, ctDNA appears to be homogenous within the plant. Most of the DNA is organized into single copy sequences, and a number of functional genes have been identified, including ribosomal DNA, genes specifying the large subunit of ribulose bisphosphate carboxylase (the main chloroplast protein), and cytochrome f. Some chloroplast proteins, for example

the small ribulose bisphosphate carboxylase subunit, are known to be specified by the nuclear genome.

The organization of the chloroplast genome is rather similar to that found in bacteria. The ribosomes (70s) and tRNAs are quite distinct from those of the cytoplasm and mitochondria, and the translation of chloroplast-produced proteins is sensitive to prokaryotic protein inhibitors such as chloramphenicol.

The contribution of ctDNA to gene expression in plant cell cultures is likely to depend upon the degree of photoautotrophy (light-dependent growth) displayed by the cells. Under usual culture conditions, consisting of relatively low light, atmospheric carbon dioxide content, and an exogenous carbon supply, plant cell cultures contain poorly differentiated green chloroplasts or non-green proplastids and are unable to fix carbon dioxide photosynthetically to any significant extent. However, in photoautotrophic or 'photomixotrophic' conditions some cell cultures exhibit green chloroplast differentiation and it can be expected that chloroplasts function under these conditions. It has been demonstrated that the non-green proplastids in photomixotrophic cultures of spinach contain lower copy numbers of plastid DNA than mature chloroplasts (approximately 1100 as opposed to approximately 6000). It is also clear that in this system there is a direct relationship between gene dosage and the level of transcripts in differentiating plastids. Chloroplasts are useful visible markers in studies of culture stability and protoplast fusion, as well as providing good genetic markers (see Chapters 3 and 5).

The Variability of Plant Cell Cultures

A glance down a light microscope at a plant cell suspension will reveal its morphological heterogeneity. Using phase contrast, cells of vastly different sizes and shapes can be easily distinguished, usually arranged into clumps of 2–100 cells. In rapidly dividing cultures, small, starchy, densely cytoplasmic cells predominate, whereas in stationary-phase cultures larger, more vacuolated cells tend to be in excess. At any stage of the culture cycle, however, a mixture of both cell types will be obvious.

This morphological variability may to some extent reflect the fact that plant cell cultures are naturally asynchronous with respect to the mitotic cycle, and at any one time a culture will contain a combination of cells in different cell cycle phases. However, the observable variation may also reflect underlying biochemical or genetic diversities. Examples of variant types which have been detected in culture, and the suggested basis for such heterogeneity, will now be discussed.

EXPRESSION OF PHENOTYPIC VARIATION IN CELL CULTURES

A few of the main groups of phenotypic variants which have been identified in plant cell cultures are described briefly below. Variability in embryogenic and regenerative potential has also frequently been reported, and this phenomenon will be dealt with in Chapter 4.

Hormone Habituation

Plant cell cultures are normally established and maintained on media containing an auxin and a cytokinin. Removal of either hormone from the medium would normally result in culture death; however, auxin- or cytokinin-independent growth has often been observed to arise during the life of a culture, sometimes in response to an induction process such as the application of hormone or high incubation temperatures. This phenomenon, which is known as *hormone habituation*, has never been adequately explained. As the molecular basis of hormone action in plant cells is poorly understood, it is only possible to speculate on the changes at a physiological or genetic level that might bring about hormone-independent growth. However, the reversible nature of habituation makes it likely that it is not genetically determined.

Resistance to Metabolic Toxins

There have been numerous reports on the isolation of plant cell lines exhibiting resistance to a wide variety of antimetabolites. No specific mutagenic treatment is necessary in order to produce such variants; they can arise spontaneously in culture. The application of a selection pressure, for example toxic amino-acid analogues, herbicides, or antibiotics, produces a temporary inhibition of growth of the culture as a whole during which only the resistant cells can divide. Possible mechanisms for resistance are wide-ranging; they include inhibition of uptake into the cell, and increases in the activities of target enzyme via either transcriptional control or a change in the copy number of the coding sequence. For further details of these variants and their application, see Chapter 5.

Secondary Metabolite Yield

A discussion of the various types of secondary metabolites which can be accumulated by plant cell cultures is presented in Chapter 6. Numerous reports of variability in product level within cell populations and in the same cell line over consecutive subcultures have been made. For example, under certain conditions cultures of many species including *Daucus carota* (wild carrot) and *Catharanthus roseus* (Madascagar periwinkle) will accumulate anthocyanins; however, the distribution of anthocyanin pigmentation throughout the culture is never homogeneous. Likewise, varying degrees of fluorescence can be observed in serpentine-producing cell cultures of *C. roseus* cultured on production media. Attempts to select high-producing cell lines by the culturing of callus or suspension subclones usually give rise to cell lines displaying a wide range of product levels. This strategy has been successfully exploited in the selection of high shikonin-producing cell lines from cultures of *Lithospermum erythrorhizon*, part of the development programme for the first commercial plant cell culture operation (see Chapter 6).

While this is not necessarily the rule, many plant cell cultures also exhibit variation in secondary product yield over successive subcultures (Table 2.2). This phenomenon obviously constitutes a potential problem for the commercialization of fine chemical production by plant cell cultures, particularly when the instability takes the form of a consistent reduction in yield. However, this process of yield deterioration can be arrested by making selections at regular intervals.

Table 2.2 Examples of plant cell suspensions exhibiting secondary metabolite yield instability over successive subcultures

Species	Natural product
Nicotiana rustica	Nicotine
Lithospermum erythrorhizon	Shikonin
Catharanthus roseus	Serpentine and ajmalicine
Daucus carota	Anthocyanins
Anchusa officinalis	Rosmarinic acid
Coptis japonica	Berberine
Digitalis lanata	β-Methyl-digoxin
Perganum harmala	Serotonin
Nicotiana tabacum	Cinnamoyl putrescines

Table 2.3 Crop species displaying somaclonal variation (from Scowcroft and Larkin, 1982)

Species	Examples of useful variant characteristics in regenerated plants
Sugar cane	Increased resistance to Fiji disease virus and downy mildew
Potato	Altered tuber morphology
Tobacco	Yield, alkaloid level
Onion	Bulb size and shape
Pelargonium	Flower and leaf morphology, anthocyanin pigmentation

SOMACLONAL VARIATION

The term *somaclonal variation* was adopted by Larkin and Scowcroft in 1981 to describe the variation occurring in plants regenerated from cultured cells or cultured plant tissues (Scowcroft and Larkin, 1982). It was suggested that the potential of tissue cultures for the induction of variation could be exploited by crop breeders, and it is now apparent from a number of examples that this variation has some value (Table 2.3). In certain cases, somaclonal variants display much improved characteristics when compared with those of the source plant from which the cultures were derived. As one example, a bioassay developed to ascertain the sensitivity of sugar-cane leaves to fungal toxins was used to demonstrate that many plants regenerated from sugar-cane tissue cultures displayed increased resistance to different fungal toxins. An important observation was that in many of the somaclones the enhanced resistance was maintained through subsequent generations, indicating that in this case the tissue-culture derived alterations were stable.

Stability of somaclones is, however, not always the rule. Desirable characteristics present in newly regenerated plants may be gradually lost over successive generations; in such cases the mechanisms determining the newly arisen phenotype are likely to be different from those responsible for stable changes.

It should be clear from the discussion above that the variation generated via the tissue culture process could be interpreted as a hindrance to progress, for example in an industrial programme aimed at drug production from cell cultures. In this situation, unpredictable yield fluctuations of 50% or more may make all the difference between an interesting observation and a commercial process. Alternatively, crop breeders may regard somaclonal variation as a positive advantage. For whatever reason, if it is ever going to be possible to exercise control over culture variation, efforts must be made to understand more extensively the mechanisms which cause it. In the past, these have been divided, perhaps somewhat artificially, into two categories: genetic and epigenetic.

THE CONCEPT OF A GENETIC BASIS FOR VARIATION

If a new characteristic can be transmitted sexually, then it is generally described as *genetic*. This infers that the variation has arisen from a structural alteration in the DNA, and is therefore a mutation, a fact not often established when elucidating the nature of culture-induced variants.

It may be suspected that a plant cell culture variant is genetic if it appears to be stable through several generations of subculture, in the absence of selection pressure, and if the culture conditions are altered in any other way. Persistent characteristics under these circumstances could be termed 'heritable', but are not necessarily genetically determined.

The investigation of culture variant stability via sexual transmission may not always be practical, as it is necessary not only to regenerate the undifferentiated cultures back to the whole plant, but also to ensure that these regenerants are fertile. This requirement highlights a problem often associated with particular species, that of recalcitrance. Plant regeneration from cell cultures of some leguminous plants, cereals, and some forest trees can be difficult, and a strong genotype effect is often observed, so that only selected cultivars will respond to the conditions known to favour regeneration in more amenable species. In addition, the loss of regenerative potential with increasing time in culture is a phenomenon commonly observed for many plant species. This topic is further dealt with in Chapter 4. Some examples of somaclonal variants which have been shown to have a genetic basis are discussed in Chapter 5.

EPIGENETIC VARIATION

Whereas it is possible to suggest numerous genetic mechanisms which could provide a basis for heritable, sexually transmitted characteristics, *epigenetic* changes pose more of a problem. This term has frequently been used to describe any culture-derived variants which exhibit some stability but cannot be shown to be genetic. However, a definition has been proposed which describes epigenetic alterations as heritable changes in gene expression which do not arise as a result of permanent alterations in the genome of the cell. The ways in which epigenetic and genetic changes differ are summarized in Table 2.4, but these criteria should only be regarded as guidelines. For instance, under the influence of transposons, some

Table 2.4 Criteria for the classification of tissue culture variants as genetic or epigenetic

Feature	*Genetic*	*Epigenetic*
Induction of variation	Low: 10^{-5}–10^{-7} per cell generation	High: $> 10^{-3}$ per cell generation
Nature of change in response to inducer	Random	Directed
Stability of change in somatic lineages	Usually stable	Stable, but reversal can occur at high rates
Transmission of change to meiotic progeny	Yes	No

mutations could arise at much higher rates than normally expected and these could erroneously be categorized as 'epigenetic'. Similarly, under non-selective conditions, the rate of induction of epigenetic variants might be very low.

Culture-induced variations of diverse types have been described as epigenetic. For example, cell lines of *Nicotiana sylvestris* have been isolated exhibiting varying degrees of chilling resistance. From these cultures, regenerants were obtained which also displayed the selected phenotype; however, the character was not transmitted to sexual progeny, indicating a probable epigenetic basis. Another example which is worth mentioning concerns the slow but persistent induction of urease levels in tobacco cell suspension cultures grown on medium containing urea rather than the usual nitrate. If urea was then replaced with nitrate as the major nitrogen source, urease declined. However, on further challenging with urea-only medium, urease induction was relatively rapid. The calculated rates of induction were high, in the region of 10^{-3} per cell generation, which again suggests a possible epigenetic mechanism.

Genetic Mechanisms for Variation

In this section selected examples of the various types of genetic alterations observed in plant cell cultures will be discussed. These have not generally been correlated with particular phenotypic changes, but any exceptions to this statement will be mentioned where appropriate.

OBSERVED CHANGES IN THE NUCLEAR GENOME

Variation in Chromosome Number
Because of the relative ease with which chromosomes can be visualized by the use of staining techniques and a light microscope, the literature abounds with reports

Table 2.5   Ploidy variation with plant tissue (from Nagl, 1978)

Species	Tissue	Maximum ploidy level
Bryonia dioica	Anther hairs	$256n$
(Cucurbitales)	Endosperm	$48n$
	Corolla hairs	$8n$
Phaseolus coccineus	Suspensor	$4{,}096n$
(Fabales)	Endosperm	$96n$
Vicia faba	Stem epidermis	$16n$
(Fabales)	Stem pith	$8n$
	Cotyledon	$16n$

of plant cell cultures displaying abnormal karyotypes (Bayliss, 1980; D'Amato, 1985). Before discussing some of these examples in more detail, it is worth putting them into context by briefly considering the situation in the whole plant.

Polyploidy occurs quite naturally *in vivo* (see Nagl, 1978), being especially common amongst angiosperms (Table 2.5). The ability of whole plants to derive a new individual from any somatic cell and the formation of new meristems means that new plant populations can arise carrying endoreduplicated genomes in the germ line. Once formed, meristems appear to be genetically stable, a state which is probably conferred by the tight control of the cycle of DNA synthesis and mitosis, and the continuity of cell division (and see pp. 87–8). The high frequency with which polyploidy has been observed in callus and suspension cultures suggests that, *in vitro*, the mechanisms which normally control the occurrence of aberrant mitotic events may operate at a reduced level.

Many different mechanisms could account for the production of polyploid cells in culture, and there is substantial evidence for their occurrence under *in vitro* conditions. For instance, synchronous division of the nuclei of binucleate cells has been observed to occur in some cultures; the nuclear spindles could fuse during this process, as has been detected in root tips after the induction of binucleate cells by treatment with caffeine. Polyploidization could also arise as a result of other reported phenomena such as chromosome lagging during anaphase, and the formation of a single nucleus after chromatid segregation due to the poor functioning of the spindle. Abnormal spindle formation is also likely to be a major cause of triploid cell production *in vitro*; it has been observed that multipolar spindles can commonly occur during the division of polyploid cells both in culture and in the whole plant.

Aneuploidy describes the status of a nucleus with a chromosome number which is not a multiple of the germ line number; for instance in a cell in which the diploid chromosome number $(2n)$ is 6, aneuploid configurations may be for example $2n-1$ (Cramer *et al.*, 1985) or $2n+2$ (Newton, 1988). Aneuploidy is frequently detected in plant cell cultures and can arise as a result of aberrant mitotic events including, for example, those which follow nuclear fragmentation. This phenomenon, which

Table 2.6 Chromosome number variation in plants regenerated from culture (from D'Amato, 1985)

Species	Chromosome number	
	Source plant	Regenerants
Asparagus officinalis	Diploid, $2n = 20$	20, 40
Avena sativa	Diploid, $2n = 42$	41, 42, 43
Brassica oleracea	Diploid, $2n = 18$	18, 36, 72
Datura innoxia	Haploid, $n = 12$	12, 24, 48
Daucus carota	Diploid, $2n = 18$	18
Haworthia setata	Diploid, $2n = 14$	14, 27, 28
Nicotiana tabacum	Diploid, $2n = 48$	65–71
Oryza sativa	Diploid, $2n = 24$	24, 48
Prunus amygdalus	Diploid, $2n = 16$	16
Solanum tuberosum	Haploid, $n = 24$	46, 48, 51–93
Triticum aestivum	Diploid, $2n = 42$	42
Zea mays	Diploid, $2n = 20$	19, 20, 21, 40

involves the furrowing of the nucleus and formation of new nuclei without any apparent nuclear spindle formation, has been often observed. For example, in one study, the nuclear events occurring during callus induction from immature cotyledons of *Vicia faba* were studied. A lobing of explant cell nuclei was followed within a few days by mitosis not only of diploid and polyploid nuclei but also of fragmented nuclei, which contained a very wide range of chromosome numbers. Other events which could lead to the production of aneuploidy such as multipolar anaphases and lagging chromosomes have already been mentioned. Non-disjunction, where both chromatids pass to one pole of the spindle instead of to opposite poles, is an additional mechanism.

Chromosome Number Variation in Regenerants
Given the many reports of chromosome number variation amongst plants regenerated from culture, polyploidy and aneuploidy might be expected to constitute a major source of 'somaclonal variation' in plant tissue culture (Table 2.6). However, it has not generally been possible to correlate these cytogenetic changes with phenotypic differences in the regenerated plants. Somaclones of tobacco, cauliflower, sugar-cane, millet, and many other plant species have been shown to exhibit apparently normal karyotypes. However, there is some evidence that the ploidy status of a culture may to some extent influence its regenerative potential.

It is obvious from Table 2.6 that regenerants do not always display unusual chromosome numbers. In view of the diverse ploidy levels detected in the cultures from which some of these plants were derived, this may seem surprising. For instance, in the *Daucus carota* suspension from which the diploid regenerants were obtained, haploid, diploid, tetraploid, and octoploid nuclei were observed. Likewise, the *Prunus amygdalus* callus which also yielded only diploid regenerants exhibited an array of aneuploid, diploid, triploid, and tetraploid cells. It would appear therefore, that in some species at least, diploid cells possess a selective advantage over other cell types during the process of regeneration.

Chromosomal Aberrations

Deficiencies, duplications, translocations, and other gross alterations in chromosome structure can only be detected by the use of chromosome banding techniques such as the Giemsa staining method, or by the investigation of meiotic events in regenerated plants. Consequently, reports of such structural changes are less widespread than those relating to ploidy variation. However, aberrations at the level of the chromatid such as chromatid deletions, interchanges, and pseudo-chiasmata have been observed as well as chromosomal irregularities such as dicentrics, rings, and so on. Such large rearrangements of the genome are likely to bring about dramatic changes in gene expression. For instance, while chromatid deletions are likely to unmask normally recessive genes, transposition of a chromosome fragment to a new position may lead to suppression or expression of genes adjacent to the site of excision or insertion.

It has been suggested that the maintenance of the diploid state in plant cell cultures is a poor measure of stability, because, without performing structural analyses, the fine structure of chromosomes with respect to rearrangements is not known. Evidence indicates that the genetic nature of cytologically 'diploid' tissue culture regenerants should likewise be in doubt until proven otherwise.

Gene Mutations

The criteria used to provide evidence for the genetic basis of cell culture variation have not always been met; as already mentioned, it is necessary to demonstrate the expression of the variant phenotype in regenerants and via sexual transmission to progeny. Many of these heritable mutations have shown a complex inheritance pattern which suggests the contribution of a number of genes, while others more clearly have a single-gene basis.

Lack of expression of the culture-derived characteristic in regenerants may not necessarily rule out a genetic basis, however. It is quite reasonable to expect the control of specific gene expression *in vitro* and *in vivo* to differ, and there are a number of cases reported in which characteristics shown by cultures are not seen in the whole plant. For instance, *N. tabacum* cell suspensions were treated with a tryptophan analogue, 5-methyl tryptophan. Variant cultures with enhanced levels of a feedback-control-resistant anthranilate synthase enzyme were selected; these accumulated around 30 times more tryptophan than wild-type cells. This characteristic was lost in the regenerated plants, but cell cultures initiated from their leaves again exhibited resistance to the analogue.

Alternative lines of evidence for true mutations would be the demonstration of an altered polypeptide sequence or gene structure, in those cases when a specific gene difference can be identified. Molecular analysis is already being used to investigate the basis of somaclonal variation; for instance, the use of labelled probes containing part of the DNA coding for 25s ribosomal DNA showed that some regenerated potato plants exhibiting morphological variation were also deficient in rDNA (Landsmann and Uhrig, 1985). A similar type of study revealed that certain T-DNA genes (see Chapter 5) in three-year-old tobacco crown gall lines were inactivated by extensive genome rearrangements, including deletions and possibly gene amplifications. More recently, somaclonal variants having single base-pair alterations have been reported, for example in the alcohol dehydrogenase gene of maize. Now that the investigation of culture-induced variation is accessible to molecular biological techniques, reports of this nature should become more common.

CONTRIBUTION OF THE CYTOPLASMIC GENOME TO VARIABILITY

There is as yet little evidence to suggest that the culture process itself contributes to variability in the cytoplasmic genome. Comparative studies made on mtDNA obtained from whole plants and their cultures, and from cultures established for long culture periods, have usually indicated stable restriction fragment patterns. However, in suspension cultures of tobacco, variability in mtDNA restriction fragment patterns has been detected, and it has also been found that in these cultures, most of the mtDNA is organized as amplified small circular DNA molecules which are derived from the main mitochondrial genome (Hanson, 1984; Newton, 1988).

Attempts have been made to select for mtDNA variants in tissue culture, though lack of information regarding the phenotypes specified by the mitochondrial genes has limited this approach. CMS in maize is currently the only mtDNA-encoded characteristic which can be readily selected in culture. Male-fertile plants have been found to be resistant to a toxin from the fungus *Helminthosporium maydis*, whereas male-sterile plants are susceptible. It has been demonstrated that callus cultures derived from toxin-sensitive plants can yield resistant cell lines after a lengthy selection in the presence of the toxin. Plants regenerated from these selected cell lines exhibited both toxin resistance and male fertility, and some plants were found to have altered mtDNA restriction fragment patterns.

The inheritance of chloroplasts and mitochondria in hybrid plants derived by fusing the protoplasts of related species has also been explored at a molecular level. Analysis of chloroplast DNA from, for example, hybrids of *N. glauca* and *N. langsdorfii*, has yielded no evidence of recombination or mutation. However, a number of studies on mtDNA inheritance in similar somatic hybrids of *Nicotiana* spp. have revealed that mtDNA recombination may occur as a result of protoplast fusion.

In summary, evidence suggests that it is usually only when selection pressures, such as toxin application, or stressful conditions, such as protoplast fusion, are imposed that alterations in the mitochondrial or chloroplast genome are likely to be brought about.

Possible Mechanisms Underlying Epigenetic Variation

Awareness is growing of a number of mechanisms which may underlie the 'epigenetic' nature of variation in cultured plant cells. While the existence of a few of these phenomena has been suspected for some years, it is only with the advent of DNA methodologies providing an increased sensitivity in analysis that a greater understanding has been reached. Three such processes will be discussed below.

GENE AMPLIFICATION

The selective amplification of genes in some animal systems has been extensively investigated; for instance, in mouse cell cultures selected for resistance to the antitumour drug methotrexate, the gene encoding the target enzyme, dihydro-folate reductase (DHFR) was found to be amplified by as much as 200-fold. The mode of selection appears to be of great importance in determining the mechanism of resistance, in that the toxin or drug must be administered to the tissue or cell culture in stepwise increasing doses. As a result of this procedure, the resistant cells survive apparently by gradual adaptation to the adverse conditions, rather than by random mutation.

Plant cell cultures selected for resistance to a range of antimetabolites often exhibit much higher target enzyme activities than normal. In some cases the enhanced activities have been attributed to enzyme over-production, and in a few instances this has been shown to be caused by gene amplification (Table 2.7). Though it has not been demonstrated, the gradual increase in urease levels observed in tobacco cells subcultured from a nitrate-containing to a urea-containing medium could also be explained by amplification of the urease gene (see p. 35).

By extrapolation from animal systems, this phenomenon can be viewed as producing relatively stable or unstable changes depending upon the location of the amplified sequences. It has been found that resistant phenotypes can be stably inherited when the amplification is located within the chromosomal body. However, if the amplified gene sequences are organized on extrachromosomal elements, then it is likely that once the selection pressure is removed, these elements will be lost from the genome over successive cell cycles. The phenotype would then revert to normal. Certainly this is one mechanism by which plant cell cultures could achieve their extensive potential for adaptation to new conditions. Though routine culture conditions tend to be considered as 'non-selective', it is impossible to predict just how much the genome might alter in such an unnatural environment. It is known that gene amplification can accompany the tissue culture process; e.g. in rice, certain nuclear DNA sequences became amplified during de-differentiation of the explant tissue.

GENE METHYLATION

Up to 7% in animal cells and often more than 25% in plant cells of the cytosine residues in eukaryotic DNA can be methylated. The function of DNA methylation is still a controversial issue, but there is now considerable evidence that the degree of DNA methylation of some genes is inversely correlated with their expression. Methylation patterns are heritable, being conferred and maintained by the action of respectively *de novo* methylase and maintenance methylase enzymes, and in

Table 2.7 Examples of gene amplification in plant cell cultures

Species	Inhibitors	Enzyme	Tested criteria[a]			
			1	2	3	4
Alfalfa	L-phosphino-thricin	Glutamine synthetase	✓	✓	✓	✓
Petunia	Glyphosate	EPSP synthase	✓	✓	✓	✓
Datura	Chlorsulfuron	Acetolactate synthase				✓

[a]After selection in the presence of herbicide the following alterations were demonstrated: 1, increased enzyme activity; 2, increased enzyme level; 3, increased mRNA level; 4, increased gene copy number

Table 2.8 Alterations in DNA methylation arising in culture

Species	Tissue type	Observed methylation change
Daucus carota (wild carrot)	Cell suspension	Altered methylation pattern of random DNA fragments during embryogenesis
Glycine max (soybean)	Intact plants Suspensions	Reduction in methylation of 5s rDNA during culture induction
Zea mays (maize)	Regenerants from embryogenic callus	Different methylation pattern of genomic DNA between regenerants

animal systems cell differentiation has been found to be accompanied by alterations in patterns of gene methylation and gene expression.

In plants, some attempts have been made to determine the relationship between gene methylation and transcriptional activity, often with no satisfactory conclusion. There are examples, however, which provide some evidence of an inverse relationship between gene expression and the extent of methylation at specific sites. For example, in maize plants it was found that HhaI restriction sites in a zein gene were methylated in ears and embryo, in which no zein expression was detected. However in the endosperm where zein accumulation occurred, these sites were hypomethylated. In a more recent study, T-DNA expression and extensive genomic demethylation in tobacco suspension cultures was brought about by treatment with 5-azacytosine derivatives, these being known DNA demethylating agents (Klaas *et al.*, 1989).

In plant cell cultures, alterations in DNA methylation patterns have been observed to occur spontaneously during the processes of differentiation or de-differentiation, and a few examples are shown in Table 2.8. The occurrence of these changes, and the evidence that there may be a role for cytosine methylation in the control of gene expression, indicate that this phenomenon should be regarded as one possible method for the production of epigenetic variation in plant cell and tissue cultures.

CONTROLLING ELEMENTS

In the 1940s Barbara McClintock proposed that an explanation for the long-term instability of maize kernel phenotype was the existence of 'controlling elements'. These were defined as genetic factors which were capable of moving around the genome and modifying gene expression, and their existence at this time was postulated entirely on the basis of genetic analysis. Since then, progress in molecular techniques has led to the demonstration of a number of controlling elements, for example the so-called Ds, Ac, and Mu elements of maize. In addition, similar factors have been found to influence flower pigment variation in *Antirrhinum majus*. It is highly likely that controlling elements are a widespread phenomenon in the plant kingdom, their recognition in maize and *Antirrhinum* having been made easier by the relatively comprehensive knowledge of the genetics of these species, and the fact that the new phenotypes produced by the effect of these elements were obvious pigmentation variants.

Generally speaking, insertion of a controlling element causes suppression of transcriptional activity at the site of insertion or the production of an aberrant transcription product, while excision may allow reversion to the wild-type phenotype. Several factors appear to affect the rate of excision, including virus infection, temperature, chromosome breakage, and the genetic background. There is now some evidence that the culture process may also affect controlling element mobility. For instance, from tissue cultures of a white-flowered mutant of alfalfa, more than 20% of the regenerants exhibited the purple-flowered phenotype (Groose and Bingham, 1986). Genetic analysis indicated that while the wild-type and mutant alleles were stable and sexually transmitted, the culture process appeared to trigger reversion, and controlling element involvement has been suggested. In a similar study, maize plants were regenerated from tissue culture and found to contain an active Ac element, whereas none had been detected in the initial explant sources. In conclusion, current evidence suggests that controlling element activity may be triggered by the culture process, and in itself may underlie some culture-derived genetic variability.

Why Plant Cell Cultures are Genetically Variable

In the sections above a number of mechanisms have been described which may contribute to the extensive genetic variability of undifferentiated cell cultures; phenomena which can sometimes be observed *in vitro*, but more often at the level of the regenerated plant or 'somaclone'. So far, however, the discussion has not dealt with the key questions of why plant cell cultures should be subject to such variation, and at what stages in the tissue culture process the variability is induced. A brief consideration of these points concludes the chapter.

Two main factors influencing culture variability should be considered at the culture initiation stage. First of all, the explant source is likely to be heterogeneous with respect to the state of differentiation, age, and ploidy status of the component cells. Callus can be induced via the division of a range of somatic cell types, depending upon the species, explant tissue, and culture conditions, and it is highly likely that cultures arising from such complex mixtures of cells are themselves mixed populations from an early stage. A process of sorting and selection may then follow, some cell types being lost from the callus line during successive subculture while others achieve a greater level of dominance. In the event of single-cell derivation of culture tissue, there still exists the possibility of inducing variability. For example, a polyploid explant cell may undergo further polyploidization or reversion to the diploid state in the early stages of culture, while cells derived from cultivars exhibiting controlling element activity may have the potential to generate insertion mutants or revertants.

In addition, the composition of the callus-inducing medium, especially with respect to the phytohormone composition, may affect culture stability. It has long been suspected that high levels (greater than 5 mg/L) of certain auxins including 2,4-D can lead to chromosomal aberrations (Bayliss, 1980). On the other hand, some hormone compositions may lead to partial or complete organogenesis. The maintenance of true meristems in 'callus' tissue may confer greater stability of phenotype, as discussed below.

VARIATION ARISING DURING CULTURE

While the nature of the explant tissue undoubtedly contributes to culture variation, substantial evidence now points to the fact that much of the variability observed in cultured material or in regenerants arises as a direct result of the culture process itself.

In one experiment (Prat, 1983) protoplasts were produced from a tobacco cultivar which had been subjected to an exhaustive programme of selfing, haploidization, and chromosome doubling, thus eliminating the possibility of residual heterozygosity. However, plants regenerated from the cultured protoplasts were shown to be heterozygous for mutations affecting a number of morphological characteristics, indicating that the culture phase generated a considerable amount of variability.

Another study, described in Fig. 2.2, made use of a known mutant of tobacco, 'sulphur'. The homozygotes and the heterozygotes were all readily recognizable on the basis of the leaf morphology; the wild-type *su/su* having normal dark green leaves, *Su/Su* having an albino phenotype, and the heterozygous *Su/su* possessing yellow-green leaves. The authors interpreted the detection of 'homogeneous variant' colonies as demonstrating variability pre-existing in the explant, while 'heterogeneous variant' colonies were only possible if genetic alterations occurred after protoplast preparation. Their data suggested that the latter type of variation occurred much more frequently than explant-derived variability (Lorz and Scowcroft, 1983).

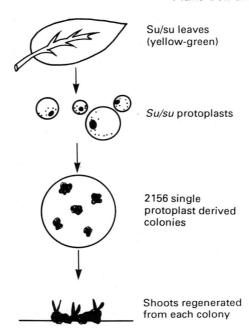

Su/su leaves
(yellow-green)

Su/su protoplasts

2156 single
protoplast derived
colonies

Shoots regenerated
from each colony

shoots analysed and the protoplast colonies classified as:

* *Parental*: all shoots derived from a colony were *Su/su*
* *Homogeneous variant*: *all* shoots derived from the colony were either *Su/su* or *su/su*
* *Heterogeneous variant*: colonies gave rise to *Su/su + su/su* shoots, or to *Su/su + Su/Su* shoots.

Fig. 2.2 Detection of variation arising during culture of tobacco protoplasts (Lorz and Scowcroft, 1983).

A further example relating to the reversion of alfalfa mutations during culture has already been mentioned (p. 42).

MERISTEM STABILITY

It has frequently been observed that the amount of variation produced in callus and suspension cultures is much greater than that to be found in axillary shoots proliferating from stem nodes. For example, while the commercial production of uniform potato plants via shoot cultures is now possible, those propagated adventitiously (i.e. via a callus stage) exhibit a high level of somaclonal variation. In *Asparagus*, plants regenerated from callus included a high proportion of polyploids, while those produced from axillary shoots were stable.

The main difference between propagation via axillary or adventitious shoots lies in the maintenance or loss of an organized meristem in the primary culture.

As already mentioned, callus may arise via cell division in a range of somatic cell types. With each successive subculture increasingly more genetic and epigenetic alterations can accumulate, which could then be transferred into regenerated shoots. However, axillary shoots derive directly from organized meristems and are subject to only low levels of variation.

Meristem stability is probably the result of a number of factors. In the first place, it has been suggested that meristems may possess more efficient DNA repair mechanisms than disorganized tissues so that mutations are less likely to accumulate. Secondly, shoot apices have a complex structure consisting of two or more multicellular layers, all of which are maintained in axillary meristems. Mutations affecting entire tissue layers are more likely to be rare events, and even then will produce only partially mutant shoots, known as *chimeras*. More frequently, mutations occurring within a meristem would only affect a small group of cells and are therefore unlikely to be maintained in subsequent axillary meristems.

When multiplying shoots via tissue culture, it is therefore evident that the particular strategy employed should be selected with great care. If the production of uniform material is the aim, than a callus stage should be avoided; however, a short callus step prior to adventitious shoot induction may be highly effective if the objective is to generate variability in a crop.

In the context of producing valuable plant metabolites from tissue culture, the avoidance of genetic variation though meristem multiplication has generated considerable interest in the technique of 'hairy root culture'. Here, the insertion of *Agrobacterium rhizogenes* R_i plasmid DNA into the plant genome has in certain cases stimulated rapid root growth. When excised, these transformed tissues can proliferate indefinitely in simple media, and most importantly can generally continue to accumulate compounds normally synthesized in the roots in a stable fashion. This technique is discussed in greater detail in Chapter 6.

THE AVOIDANCE OF TISSUE CULTURE VARIATION

In summary, tissue-culture-induced or 'somaclonal' variation has an intrinsic value to the plant breeder, providing an alternative source of genetic variability for selection. However, in those areas of plant tissue in which the *maintenance* of a particular phenotype is of paramount importance, i.e. micropropagation or large-scale culture for metabolite production, thorough preparation prior to initiating cultures may well reap rewards in the long-term. Questions which should be considered are:

- What type of culture is the objective? For micropropagation purposes, a meristem should be maintained if possible. However, large-scale plant culture technology is at present based upon the use of undifferentiated cell suspension cultures. Differentiated culture systems (e.g. shoot, root) are now being more extensively researched for metabolite production, and the technology for handling these systems may well develop rapidly.
- If an undifferentiated culture system is the objective, is a range of explant

sources available? The cultivar may be of particular importance; for instance, recently hybridized material may generate relatively more variation in culture, and transposable element activity may be at a higher level in certain varieties of species known to harbour them. When initiating new cultures it is therefore always important to utilize explant material of diverse genetic backgrounds.

- Can the environment be controlled? The maintenance of constant light, temperature, and humidity is of critical importance. Equally, suspension cultures should be cultured under uniform aeration and agitation. The subculture regime should be standardized and preferably always performed by the same person.

It should be obvious from the discussion above that our understanding of the basis of tissue culture variability is still very poor. The reasons for its occurrence can only be speculated upon, and though certain precautions can be taken we are certainly no nearer to controlling it. This account is intended only as a summary of the phenomenon in the context of biotechnological processes, and for further details the references below are recommended.

References and Further Reading

Bayliss, M.W. (1980). Chromosomal variation in plant tissues in culture. In *Perspectives in Plant Cell and Tissue Culture* (I.K. Vasil, ed.) *International Review of Cytology Supplement* **11A**, Academic Press, London, pp. 113–144.

Bayliss, M.W. (1985). Regulation of the cell division cycle in cultured plant cells. In *The Cell Division Cycle in Plants* (J.A. Bryant and D. Francis, eds). Cambridge University Press, Cambridge, pp. 157–177.

Cramer, C.L., Bell, J.N., Ryder, T.B. *et al.* (1985). Co-ordinated synthesis of phytoalexin biosynthetic enzymes in biologically-stressed cells of bean (*Phaseolus vulgaris* L.) *EMBO Journal* **4**, 285–289.

Cramer, C.L., Edwards, K., Dron, M. *et al.* (1989). Phenylalanine ammonia-lyase gene organisation and structure. *Plant Molecular Biology* **12**, 367–385.

D'Amato, F. (1985). Cytogenetics of plant cell and tissue cutlures and their regenerants. *CRC Critical Reviews in Plant Sciences* **3**, 73–112.

Groose, R.W. and Bingham, E.T. (1986). An unstable anthocyanin mutation recovered from tissue culture of alfalfa (*Medicago sativa*). 1. High frequency of reversion upon reculture. *Plant Cell Reports* **5**, 104–107.

Hanson, M.R. (1984). Stability, variation and recombination in plant mitochondrial genomes via tissue culture and somatic hybridisation. In *Oxford Surveys of Plant Molecular and Cell Biology* Vol. 1 (B.J. Miflin, ed.) Oxford University Press, Oxford, pp.33–52.

Klaas, M., John, M.M., Crowell, D.N., and Amasino, R.M. (1989). Rapid induction of genetic demethylation and T-DNA gene expression in plant cells by 5-azacytosine derivatives. *Plant Molecular Biology* **12**, 413–423.

Landsmann, J. and Uhrig, H. (1985). Somaclonal variation in *Solanum tuberosum* detected at the molecular level. *Theoretical and Applied Genetics* **71**, 500–505.

Leaver, C.J., Boulter, D., and and Flavell, R.B. (eds) (1986). *Differential Gene Expression and Plant Development*. The Royal Society, London.

Lorz, H. and Scowcroft, W.R. (1983). Variability among plants and their progeny regenerated from protoplasts of *Su/su* heterozygotes of *Nicotiana tabacum*. *Theoretical and Applied Genetics* **66**, 67–75.

Makaroff, C. A. and Palmer, J. D. (1987). Extensive mitochondrial specific transcription of the *Brassica campestris* mitochondrial genome. *Nucleic Acids Research* **15**, 5141–5156.

Mantell, S.H., Matthews, J.A., and McKee, R.A. (1985). *Principles of Plant Biotechnology: An Introduction to Genetic Engineering in Plants*. Blackwell Scientific Publications, Oxford.

Nagl, W. (1978). *Endopolyploidy and Polyteny in Differentiation and Evolution*. North-Holland, Amsterdam.

Newton, K.J. (1988). Plant mitochondrial genomes: organization, expression and variation. *Annual Review of Plant Physiology and Plant Molecular Biology* **39**, 503–532.

Pang, P.P. and Meyerowitz, E.M. (1987). *Arabidopsis thaliana*: a model system for plant molecular biology. *Biotechnology* **5**, 1177–1181.

Prat, D. (1983). Genetic variability induced in *Nicotiana sylvestris* by protoplast culture. *Theoretical and Applied Genetics* **66**, 223–230.

Scowcroft, W.R. and Larkin, P.J. (1982). Somaclonal variation: a new option for plant improvement. In *Plant Improvement and Somatic Cell Genetics* (I.K. Vasil, W.R. Scowcroft, and K.J. Frey, eds). Academic Press, London.

Xiao, W., Saxena, P.K., King, J., and Rank, G.H. (1987). A transient duplication of the acetolactate synthase gene in a cell culture of *Datura innoxia*. *Theoretical and Applied Genetics* **74**, 417–422.

Yeoman, M.M. (ed.) (1986). *Plant Cell Culture Technology*. Blackwell Scientific Publications, Oxford.

Chapter 3

Protoplast Isolation and Fusion

GRAHAM WARREN

Introduction

One of the most striking characteristics of plant cells is the presence of a thick and relatively rigid cell wall. This wall functions in the mechanical support of the cell and in defence against physical damage and attack by pathogens. It may also play more subtle roles in cellular communication.

Unfortunately for the plant cell biologist, the cell wall also prevents direct access to the cell membrane and therefore impedes many of the manipulations we may wish to perform. For example, the study of membrane function through the use of molecular probes such as antibodies or lectins is not possible directly on plant cells. Biotechnological operations such as cell–cell fusion to produce useful hybrids are similarly precluded. An additional problem resulting from the presence of the cell wall is that single cells, as opposed to multicellular tissue fragments or cultured cell clusters, are difficult to obtain. The wall components and intercellular polymers effectively cement cells together. Because of this, single-cell cloning or selection of desirable cell types can be difficult.

A crucial step forward was made in 1960 when E.C. Cocking found that cellulose-degrading enzymes, isolated from wood-rotting fungi, were able to dissolve away the cell wall and yield the fragile but still viable *protoplast*, a cell surrounded only by its membrane. In fact protoplasts had been isolated before this time; for example, during the chopping-up of plant material some intact protoplasts escape through the ruptured walls, and protoplasts can occur naturally, e.g. in some types of endosperm. But the great power of the enzymic isolation technique was that it allowed for the first time the isolation of protoplasts in large numbers, suitable for further experimentation.

Following the first isolation of protoplasts, other advances in technique came quickly. In the 1960s, by imposition of the appropriate nutritional conditions, protoplasts were induced to divide and this line of research culminated in the full regeneration sequence of protoplast through to whole plant that was reported in 1971. This latter result was of fundamental importance to biotechnology, for it was now proven that 'improved' plant species could in principle be recovered from single protoplasts, following the appropriate manipulation or selection.

The protoplast therefore appeared to offer the ideal experimental material for a range of studies in plant cell biology and biotechnology. However, as more became known about the structure and properties of protoplasts, the initial enthusiasm diminished, tempered by a recognition of the formidable difficulties encountered in their use. Furthermore, doubts were raised concerning the suitability of protoplasts as models for the intact plant cell, a vital consideration for fundamental research into plant biology.

However, nearly 30 years of protoplast research have led us to the position where a realistic view of the potential and limitations of protoplasts is possible. Some notable achievements have already been made, some areas of early promise remain unfulfilled, but most researchers agree that protoplasts will assume an important role in the investigation of the biology of plant cells and in their biotechnological exploitation.

The Structure of Plant Protoplasts

A protoplast has been formally defined as all the constituents of a plant cell except the cell wall. The term can therefore be applied to a cell without its wall, or to the membrane-plus-cytoplasm system existing inside the intact wall. For our purposes the proper definition is the former, namely a cell lacking a wall, or more appropriately in the present context, a plant cell from which the wall has been removed. This distinction is in fact vital because we cannot assume that the removal process does not affect the properties of the protoplast, as will be discussed later. However, the protoplast is essentially a fully viable plant cell. This has been demonstrated dramatically by the successful growth (regeneration) of single protoplasts into mature, fertile plants. This total regeneration has so far been demonstrated in only a very restricted number of species, mostly (for unknown reasons) from the family *Solanaceae*. Nonetheless the list of species that can be fully regenerated is growing all the time, and most scientists working in this area believe that regeneration will prove to be a universal phenomenon.

If a protoplast is released from a cell into water, or into a medium of the same osmolarity as intercellular sap, it will burst. This simple experiment shows that in the intact cell, the cell wall restrains the expansion of the protoplast (turgor pressure), or in other words, the protoplast is usually 'pumped up' against the inside of the wall. For this reason, before protoplasts can be successfully released, the osmotic value of the bathing medium must be raised to balance, or slightly

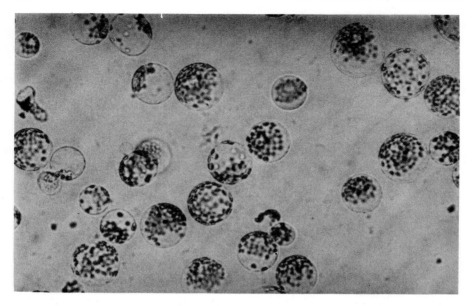

Fig. 3.1 A preparation of mesophyll protoplasts from sugar beet. Photograph kindly supplied by C. Eady. Protoplast diameter approximately 30 µm.

exceed, that of the cell's cytosol. In the latter case the protoplast (inside the wall) will shrink. This osmotic shrinkage can easily be demonstrated by placing a piece of tissue, or some cultured cells, into hyperosmotic medium. When the turgor pressure of the cell is exceeded by a uniform osmotic pressure, the protoplast adopts its characteristic spherical appearance. With the protective influence of the wall removed, the protoplast is very fragile and can easily be damaged by unfavourable changes in osmotic pressure, or by mechanical stresses. For this reason, experimental manipulations with protoplasts must always be performed with the utmost care. For example, protoplasts can be ruptured by fast pipetting through a narrow bore, or by excessive centrifugation. Even handling that causes no visible damage can result in loss of protoplast viability, as judged by impaired division or plant regeneration. In general, larger protoplasts, for example those isolated from mesophyll cells, are more fragile than those derived from smaller cell types, for example from suspension cultures, although factors other than size may be involved.

THE PROTOPLAST MEMBRANE

Since the time that protoplasts were first isolated, it has been envisaged that they would provide a unique system for the study of membrane structure and function by providing a good model for the membrane in the intact cell. However, the protoplast membrane has been found to exhibit a number of peculiarities and consequently caution must be exercised when extrapolating from results obtained with protoplasts to functions of the cellular membrane. For example, protoplast membranes usually show a large number of spherical projections (not seen in the cellular membrane) which may represent vacuoles evaginating weak regions of the lipid bilayer. It is supposed that these weak areas result in some (unknown) way from the protoplast isolation process.

Carbohydrate-binding proteins (lectins) have been used extensively to probe the surface topography and function of glycoproteins and glycolipids in animal cells. A large proportion of membrane proteins are glycosylated both in animals and plants. Glycosylation is thought to be important in protein export, stabilization against proteolysis, and cell–cell recognition events (e.g. pollen–stigma compatibility/incompatibility reactions). Many studies using lectins have been performed on plant protoplasts and have provided evidence that glycoproteins can diffuse laterally in plant membranes, giving support to the generality of the fluid-mosaic model of biological membranes. A wide range of lectins is now known, exhibiting a variety of sugar-binding specificities. Each lectin usually shows highly specific binding to one, or a very small number, of related sugars. However, it has been reported that lectins cannot distinguish between (i.e. show preferential agglutination of) protoplasts derived from different species, or between protoplasts in a given heterogeneous preparation. Thus the promise of lectins as tools in the typing of plant cells (as in human blood grouping) and in the selection of hybrid cells following cell fusion, has not been fulfilled. However, it is possible that with increased sophistication of experimental techniques and the use of lectins specific for defined sugar sequences rather than individual sugars, such cell-type-specific differences may yet be found.

When considering the surface of the membranes of isolated protoplasts, it must be remembered that the polysaccharide-hydrolysing enzymes (or other contaminating enzymes) used in protoplast isolation may modify the sugar sequences of glycoproteins or glycolipids present, thus creating an artefactual surface.

Isolated protoplast membranes are negatively charged relative to the surrounding medium. This charge causes electrostatic repulsion between adjacent protoplasts and between protoplasts and other negatively charged particles, for example certain viruses and DNA. This electrostatic force of repulsion must be overcome before protoplasts can be fused together to produce useful hybrids, and before the protoplasts can be induced to take up negatively charged macromolecules.

Protoplast Isolation

Protoplast isolation is in principle very simple; the plant cell wall is removed using polysaccharide-hydrolysing enzymes, and the protoplast is released into a medium in which the restraining effect of the cell wall is replaced by the appropriate osmotic pressure. In practice, however, the existence of a large number of inter-related variable factors associated with the starting plant material and the conditions of enzyme digestion results in a situation in which the preparation of viable protoplasts can be poorly reproducible and is devoid of a satisfactory theoretical basis. As a consequence of this, the conditions necessary for protoplast isolation from a particular tissue source cannot be predicted; instead, they must be arrived at empirically. Conditions suitable for a particular cultivar may be inappropriate for a closely related variety. Similarly, isolation parameters usually differ for cultured cells depending on factors such as composition of the growth medium and phase of the growth cycle. The ease of protoplast isolation, even under the current optimal conditions, varies enormously from species to species. Some sources produce consistently good protoplasts in high yield, and the isolation methods used are tolerant of a range of tissue growth conditions. In other cases, it is all but impossible to obtain any viable protoplasts. This latter situation is fortunately now quite rare.

PROTOPLASTING ENZYMES

The plant cell wall has three main polysaccharide components; cellulose, a β-1–4 glucan, and hemicellulose and pectin which are more complex carbohydrates. Accordingly, most protoplast isolation procedures include a cellulase, hemicellulase, and pectinase, or a crude enzyme preparation containing all three activities. Protoplasting enzymes are isolated from a variety of sources, usually fungi such as *Aspergillus niger* or *Trichoderma viride*, but an increasing number of enzymes are available, all showing slightly different substrate specificities, from more exotic sources such as snail gut. These latter enzymes can sometimes be used successfully with otherwise intractable species. Much of the uncertainty in protoplast isolation is due to possibly quite subtle differences in wall structure between species, and between plants or cells grown under different conditions. Unfortunately the elucidation of precise wall configurations is an extremely difficult biochemical problem.

Protoplasting enzymes are rarely analytically pure. In many cases purification would make their cost prohibitively high, because considerable quantities are used. Moreover, it is often a mixture of activities (not all of which have been identified), rather than single defined enzymes, that prove successful in protoplast release. However, not all the contaminants in a typical protoplasting enzyme preparation are beneficial to protoplast release and survival. The preparation may also contain ribonucleases, proteases, lipases, phenolics, and salts. These substances may reduce the yield of protoplasts, and may impair the ability of the protoplasts to divide and reform colonies and whole plants. For this reason, some workers prefer to purify their enzymes partially, for example by gel filtration, but in many cases, especially with the more amenable species, this is not usually necessary. What is important is to minimize the contact time of the plant material with enzyme solution commensurate with protoplast release, in order to minimize the damage caused by these unwanted contaminants. Incubation times for isolation of protoplasts vary from one hour to overnight, with four hours being a typical figure.

Enzymes added to plant material do not necessarily have direct access to the cell wall. The tissue cuticle and epidermis can constitute diffusion barriers to molecules of the size of proteins, as can the tightly woven nature of the wall itself. Thus penetration of protoplasting enzymes, especially into internal regions of tissue samples, can be slow. In an attempt to minimize these problems, tissue can be manually stripped of epidermis prior to protoplasting. It is also possible to subject the tissue and enzyme solution to a vacuum (vacuum infiltration) to facilitate permeation of the enzyme solution into intercellular spaces and thus speed up protoplast formation.

THE OSMOTICUM

The required osmolarity of the isolation medium, necessary for the stabilization of protoplasts during release, varies with the source of tissue used. For example, an osmotic gradient exists along the apical sections of maize roots and this situation is reflected in the osmolarity of the protoplast isolation medium required, which varies from about 0.2–0.5 M, when using tissue from these different root sections. If plasmolysis is too severe, detrimental stress effects may jeopardize the viability of the protoplast. Typical behaviour seen in plasmolysed cells includes the appearance of membrane inclusions and lipid droplets, increased nuclease activity, a decrease in the number of polyribosomes and a reduced amino acid uptake. Some of these factors are linked to decreased protein synthesis, and there is a general decrease in metabolic activity. How plasmolysis induces these effects is largely unknown. Plasmolysis can also cause uptake of molecules (including the potentially harmful contaminants in enzyme solutions) from the surrounding medium because membrane permeability can be transiently increased, for example by rupture of plasmodesmata. For this reason, in the case of particularly sensitive cells, plasmolysis may be imposed as a separate step before exposure to the enzyme solution, to limit the uptake of unwanted substances. Leakage of key metabolites can occur during plasmolysis, which in extreme cases may reduce protoplast survival rates.

The required osmotic value for the isolation medium is usually achieved by using a non-metabolized (or slowly metabolized) sugar alcohol, often sorbitol or mannitol. Sugars such as sucrose have been used, but as they are taken up by the cells their effective concentration falls. This situation may cause problems with protoplast stabilization, but it has also been exploited to provide the progressively declining osmoticum value that many protoplast types require as they undergo division and colony formation. Salt solutions have also occasionally been used as osmotica.

PRETREATMENT OF PLANT MATERIAL

The condition of the material from which protoplasts are to be isolated is critical to the success of the procedure. The factors that have so far been recognized as important in this respect include temperature, lighting, humidity and tissue age. At present the selection of the appropriate growth conditions for tissue to be protoplasted is largely a matter of trial and error because the specific cellular variables involved are poorly understood. However, many of these conditions appear to favour rapid cell division. This may confer on the derived protoplasts (by unknown mechanisms) a high potential for further division (which in turn may aid regeneration), and may also result in a reduction in the thickness of the cell walls such that protoplasts are more easily released. Other treatments designed to achieve this latter effect are the imposition of a period of dark growth on whole plant material, or growth of cultures on a medium containing reduced carbohydrate, to decrease the intracellular sugar available for wall synthesis.

A typical conditioning routine for a plant cell culture prior to protoplast release is:

- Transfer of cells to a rapid (every 7 days) subculture regime
- Increase in auxin level (favours cell division)
- Reduction of carbohydrate in the growth medium.

For whole plant material, conditioning could be achieved by:

- Frequent addition of fertilizer
- Transfer to dark 24 h before protoplasting.

The age of the plant is usually important to the success of protoplast isolation. It has been reported that tobacco plants must be taken between 40 and 60 days after germination for optimum protoplast survival. Protoplast yield and viability are (for unknown reasons) frequently low if plant material is used after flowering. As a move towards tissue standardization in terms of developmental state, the youngest fully expanded leaves on a plant are usually chosen. Because of the marked influence of the environmental conditions experienced by the parent plant on the success of protoplast isolation and survival, starting material from shoot cultures is being increasingly used. Shoot cultures, which are maintained on defined media and propagated at precise intervals, allow the growth of plant material under controlled conditions.

PROTOPLAST PURIFICATION

Once formed, healthy protoplasts need to be purified by removal of the enzyme solution, tissue debris, partially digested cells, and senescing protoplasts. Dead and dying protoplasts and cells release substances such as hydrolytic enzymes and phenolics (which can bind to and inactivate enzymes) that can prejudice the continued survival of the viable protoplasts. The overriding consideration during purification is the fragility of the protoplasts. Accordingly centrifugation (washing) steps must be performed at the lowest g force that will allow pelleting, and pipetting must be done slowly using wide-bore pipettes.

The usual methods of protoplast purification are differential centrifugation, and filtration through mesh of the appropriate pore size. It is often possible to choose a centrifugation medium of a density such that viable protoplasts float whereas dead protoplasts and cell debris are pelleted. For example, tobacco mesophyll protoplasts can be purified in this way using a 23% sucrose solution. Protoplast preparations from suspension cultures usually contain little or no tissue debris and can often be purified merely by sieving. This procedure removes undigested cell clusters and provides protoplasts of relatively uniform diameter. Following the initial purification, soluble contaminants are removed by repeated cycles of centrifugation and resuspension in fresh medium of the appropriate osmotic value.

A TYPICAL PROTOPLAST ISOLATION PROTOCOL

The following is the outline of a method used to prepare protoplasts from tobacco leaves.

- Select young, fully expanded leaves from a 40–60-day-old plant
- Cut approximately 0.5 g leaf tissue (avoiding ribs, which do not form protoplasts) into 2 mm strips using a scalpel
- Float the strips on an enzyme solution containing 1% (w/v) cellulase, 0.5% hemicellulase, 0.5% pectinase, and 0.4 M sorbitol buffered at pH 7.5
- Incubate statically overnight at 25 °C
- Gently shake the suspension to disperse digested tissue
- Filter through 50 μm nylon mesh
- Centrifuge filtrate at 100 g for 5 min
- Disperse pellet in 3 ml 23% sucrose solution and centrifuge for 5 min at 100 g
- Remove floating layer (required protoplasts) and wash three times (by centrifugation) in 0.4 M sorbitol solution pH 7.5.

TISSUE ORIGIN AND PROTOPLAST RELEASE

The most common plant tissue used as a source of protoplasts is the leaf. This material is available in relatively large quantities and contains cells that will still divide. This latter point is probably related to the relative ease with which mesophyll protoplasts divide when cultured. However, protoplasts from other sources can be of special interest, for example from pollen for the production of haploid protoplasts, or protoplasts from specialized tissues for cell biological studies. In these cases, protoplasting often turns out to be more difficult than is the case with mesophyll tissue. Frequently, much more concentrated enzyme solutions or longer incubation times are required. For example, to prepare protoplasts from elm pollen it was found necessary to incubate in cellulase for 30 h. Such harsh conditions will certainly impair the survival of many of the protoplasts formed. In fact, reports of division in protoplasts from non-mesophyll whole-plant tissue are quite rare.

There can be a tendency to think of protoplasts as a homogeneous or somehow standardized experimental material. This view is encouraged by the uniform spherical appearance of purified protoplasts. However, it is important to realize that potentially all the cell types present in the starting material can be represented in the protoplast preparation. Sometimes to minimize heterogeneity it is possible to choose plant material that contains a small number of different types of cell, possibly of clonal origin (e.g. artichoke tuber tissue). However, because not all cell types form protoplasts with the same degree of ease under a given set of conditions, a certain amount of selection for cell types can occur during protoplast isolation. For example, epidermal cells are extremely resistant to protoplasting. The whole question of the range of cell types existing in a protoplast sample remains as yet largely unresolved because the methods of cell typing are very rudimentary in plant systems, in contrast to the situation with animal cells. In a small number of cases (e.g. aleurone cells, pollen grains, root nodules, and stomatal guard cells) it has proved possible to isolate protoplasts representing essentially a single cell type. This has usually been achieved by one of three methods:

- Suitable choice of starting material so as to contain only the cells in question (e.g. pollen)
- Determination of the conditions under which one cell type will yield protoplasts preferentially (e.g. stomatal guard cell protoplasts can be isolated in very dilute enzyme solutions)
- Separation of a single protoplast type from an enzymic digest containing a mixed population (e.g. bundle sheath protoplasts have a characteristic density and can be isolated on a density gradient).

Protoplasts from single cell types will undoubtedly prove to be very useful in tackling a range of fundamental cell biological problems such as the location of specific biochemical activities and the mechanism of cell–cell interactions at the molecular level.

Metabolism of the Isolated Protoplast

Protoplasts are plant cells under stress. This stress results mainly from the osmotic conditions required for protoplast isolation, and the associated uptake of components from, and leakage of metabolites into, the isolation medium. Many of these effects are seen in intact tissues exposed to plasmolysing conditions, and so the observed behaviour is largely a response to the osmotic stress rather than to the removal of the wall as such.

Once freed from the isolation enzymes, and placed in an appropriate nutrient growth medium, the protoplast initiates the processes of adaptation and repair. Little is known about the adaptation of cells or protoplasts to culture conditions, but clearly they find themselves in a very different environment to that in the intact tissue. In general, protoplasts derived from cultured cells undergo their first divisions much more quickly than those from whole plant material. This phenomenon probably reflects the existing degree of adaptation to the culture environment in cultured cells. Soon after the initiation of the culture process there is a dramatic rise in the level of protein synthesis. In part this represents the synthesis of stress-related proteins including ubiquitin which plays a role in the recognition and degradation of damaged proteins. The major repair function, however, is the resynthesis of a cell wall.

Although there is as yet no known mechanism linking cell wall synthesis to cell division, and no theoretical reason why protoplasts should not divide in the continued absence of a wall, this latter situation is virtually never encountered.

The synthesis of a new wall occurs after a lag period during which some of the other stress-related repair/adaptation functions are carried out. This point is emphasized by the following observation. If the newly synthesized wall of a protoplast is itself removed by cellulase enzymes (no change in osmoticum now required) then wall resynthesis subsequently occurs without a lag period. Examples of typical lag periods prior to wall resynthesis are 1.5 h for leek meristem protoplasts, 3 to 24 h for tobacco, and 4 days for legumes.

The new cellulose components produced during wall resynthesis are formed as shorter-than-normal fragments that are subsequently ligated to their usual length prior to the first cell division. Matrix polysaccharides (e.g. hemicellulose) are at first lost from the protoplast and appear in the culture medium. This may be because there is an insufficient number of cellulose microfibrils present initially to anchor them. However, cell wall material is also lost in relatively large amounts from adapted suspension-cultured cells, and this behaviour can cause problems in the blocking-up of vessels used to grow plant cells on a large scale. In fact, recovery of wall polysaccharides from culture filtrates has proved to be a useful system for the study of the structure of wall components. It has been found that the structure of the newly synthesized wall is not necessarily identical to that of the parent cell, however.

A commonly observed phenomenon in preparations of cultured protoplasts is that of budding. This can be mistaken for cell division, but on closer inspection it is found to be the cell membrane bulging out from a localized area of the wall. It is thought that such local weak areas of the wall occur due to loss to the surrounding medium of matrix componenents as already described. This results in a reduced degree of cross-linking between adjacent cellulose microfibrils which in turn allows the membrane to push them apart and squeeze through.

Protoplast Culture and Regeneration

That an isolated plant cell can be induced to undergo a sustained series of divisions in the alien environment of the culture flask is a remarkable phenomenon. Even more remarkable is the fact that the same is true of fragile isolated protoplasts. Add to this the ability of dividing cells of certain species in culture to begin to coordinate these divisions and form first an embryo-like structure and ultimately a whole, fertile plant, and one can understand the excitement of the first investigators who discovered this sequence of events. The process of growth from a protoplast, cultured cell or tissue through to a plant is termed *regeneration* (Table 3.1). This should not be confused with wall regeneration as discussed earlier.

The ability of a single cell or protoplast to divide and eventually develop into a mature plant is important from two perspectives. First, it elegantly demonstrates that certain single cells (and by extrapolation possibly all cells) have all the genetic information required to specify and elaborate the whole plant. This has implications in the study of cell differentiation. Second, it presents the biotechnologist with a tool of the utmost power for the creation of new plant species or varieties starting from selected or engineered single cells. These latter topics are considered in Chapter 4.

Table 3.1 Important crop plant species that can be regenerated from protoplasts

Alfalfa	P	*Medicago sativa*
Asparagus	C	*Asparagus officinalis*
Barley	C	*Hordeum vulgare*
Carrot	P	*Daucus carota*
Clover	P[1]	*Trifolium* spp.
Cucumber	P	*Cucumis sativus*
Endive	C	*Cinchorium endivum*
Lettuce	P	*Latuca sativa*
Maize[2]	C	*Zea mays*
Millet	C	*Pennisetum americanum*
Orange	P	*Citrus sinensis*
Poplar	C	*Populus* spp.
Potato	C	*Solanum tuberosum*
Rape	P	*Brassica napus*
Rice	C	*Oryza sativa*
Strawberry	C	*Fragaria chiloensis*
Sugar beet	C	*Beta vulgaris*
Sunflower	C	*Helianthus annuus*
Tobacco	P	*Nicotiana tabacum*

P, Species will regenerate from whole plant protoplasts; C, Species will only regenerate from suspension culture-derived protoplasts.
[1] Ability to regenerate from whole plant protoplasts depends on species.
[2] Only a very few cultivars.

BASIC CONSIDERATIONS IN PROTOPLAST CULTURE

The basic requirements for the culture of protoplasts are similar to those of whole cells as considered in Chapter 1. Culture media for cells and protoplasts generally contain the same classes of component (e.g. carbon source, inorganic ions, etc.) and are often based on the same standard formulations. The essential difference is the initial inclusion in protoplast culture media of an osmoticum. A second difference is that protoplasts are frequently more demanding metabolically (i.e. have more metabolic pathways impaired) than whole cells, and it is often necessary to include a wider range of organic components or complex nutrients. Protoplasts tend also to be more sensitive to damage by certain media components or environmental conditions. As an example, agar, which is commonly used to solidify cell culture media, can be toxic to protoplasts due to impurities and sulphated components. Highly purified agarose frequently allows a higher survival rate.

TYPICAL BEHAVIOUR OF A CULTURED PROTOPLAST

Observation of the initial sequence of events occurring during the attempted culture of a protoplast preparation usually gives strong indications as to the

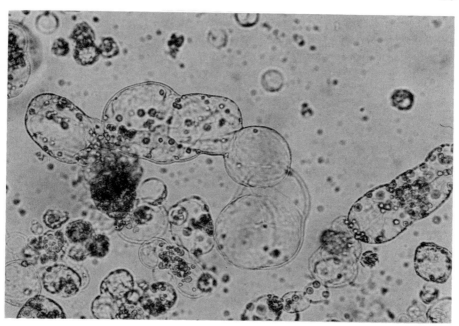

Fig. 3.2 The initial divisions of isolated protoplasts. Photograph kindly supplied by K. Lindsey.

likelihood of a successful outcome of the experiment. The first visible clues that the regeneration process is proceeding are departure from the perfectly spherical protoplast morphology due to deposition of wall material, and a marked swelling of the cells, probably reflecting increased metabolic activity (Fig. 3.2). The presence of wall material can be confirmed by staining with a polysaccharide-specific dye such as Calcofluor White. These events are typically observed from 2 to 4 days after initiation of the culture process, depending on species, and the first cell divisions occur shortly afterwards. Cultures usually contain viable and dead protoplasts. Appreciable quantities of the latter are detrimental to the continued survival of the viable protoplasts, as previously described. It is important to develop isolation and purification protocols that minimize the number of non-viable protoplasts present. If large numbers of initially healthy protoplasts die during the culture process, it is sometimes possible to rescue the remaining living ones by dilution with fresh culture medium to reduce the toxic effects of the released products. Colonies of cells originating from single protoplasts are usually visible to the naked eye within three to six weeks, and these may be removed and cultured individually if necessary. Individual colonies are likely to be clonal, but this cannot be guaranteed unless repeated reprotoplasting and replating is undertaken.

FEEDER OR NURSE CULTURES

In many of the situations in which we encounter protoplasts, it is useful to culture them either singly (for true cell cloning) or in small numbers (as is the case following selection or engineering of rare genotypes). In common with cultured cells, protoplasts exhibit a minimum density requirement for survival (Chapter 1). As a rule of thumb this density is often around 10^4 protoplasts/mL, which is far too high for conventional cloning strategies. In order to reduce this figure to a workable value we have two options. First the culture volume can be reduced, so that smaller numbers of protoplasts can be handled. Culture volumes as low as 1 µL have been used, in which as few as ten individual protoplasts can be successfully cultured. Second, it is possible to use a feeder culture to supply the appropriate levels of nutrients that the small protoplast population cannot itself sustain (Fig. 3.3). The feeder culture must be physically separate from the protoplasts to allow subsequent isolation of the protoplast-divided cells. A feeder culture normally takes the form of actively dividing cells separated from the protoplasts by a porous membrane or filter. A variation of this procedure is the use of feeder cells actually mixed with the protoplasts, but possessing some pronounced physical character, for example pigmentation, to allow subsequent isolation of the protoplast-derived cells. Irradiated feeder cells, which have lost their ability to divide, have also been used in this respect. It is also possible to use culture filtrates to supply the levels of nutrients required.

The converse of minimum density requirements, namely an upper limit on the desired concentration of protoplasts in culture experiments, is also observed. Generally densities of greater than 10^6 mL are not used because the concentration of damaged protoplasts, even in good preparations, can be high enough to have detrimental effects. In addition competition effects for available nutrients may occur.

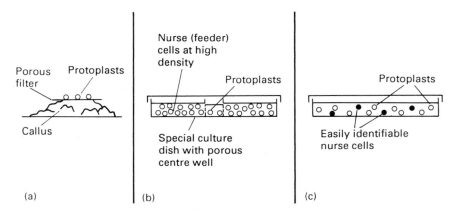

Fig. 3.3 Common types of nurse culture. (a) Callus feeder with porous membrane barrier; (b) centre well culture dish; (c) the use of morphologically distinct nurse cells.

CHANGING METABOLIC DEMANDS OF PROTOPLASTS IN CULTURE

The progressive repair, adaptation, and division of cultured protoplasts is reflected in their changing demands in terms of nutrients and environmental growth conditions. Accordingly, growth media and conditions need to be progressively modified if optimum protoplast survival is to be achieved. It has already been mentioned that lower osmotic values of growth media are required as the culture process continues. This effect is achieved either by a series of dilutions of the protoplast medium with growth medium lacking osmoticum, or by utilization of a metabolizable component of the osmoticum (e.g. glucose) such that the osmotic value of the medium automatically falls as the protoplasts grow. There is often a differential requirement for auxin during the culture process. This can take the form of an initial requirement for high auxin (possibly to induce division) following which lower levels are more conducive to sustained growth. In other cases (e.g. cereals) a progressive increase in auxin level is required. Protoplast culture is usually started on a liquid medium to facilitate these subsequent media changes.

High light intensity is frequently reported to be inhibitory to the initial divisions of mesophyll protoplasts. It is therefore usual to commence culture in the dark and gradually increase the light intensity. As with most culture requirements, the cellular basis of this effect is unknown.

PLATING EFFICIENCY

The relative merits of different culture protocols are expressed in terms of the associated plating efficiencies. There are various ways of computing this percentage, and these differences must be understood before various protocols can be compared. At the most basic level, a plating efficiency is the percentage of those protoplasts plated out on to culture medium that give rise to viable cell colonies. Some of the variations on this that are encountered include the percentage of viable (as opposed to total) plated protoplasts that give rise to cell colonies, and the percentage of the number of protoplasts undergoing the first division that subsequently form colonies. The ultimate expression of plating efficiency, however, is the percentage of plated protoplasts that give rise to plantlets.

A TYPICAL PROTOPLAST CULTURE PROTOCOL

- Resuspend washed protoplasts in liquid culture medium A (below) to give a final concentration of 10^5 protoplasts/mL
- Disperse protoplasts (500 µL aliquots) into individual wells in multiwell plates
- Seal plates with cling film to reduce evaporation
- Incubate in the dark at 25 ± 1 °C for 3–7 days
- Examine protoplasts. If divisions are observed, dilute each well with 500 µL of medium B
- Repeat the above dilution every 7 days for 3 weeks

- Transfer protoplast suspension on to medium B solidified with agar or agarose. Incubate under low light
- When colonies are large enough (approximately 2 mm diameter; around 8 weeks growth) transfer individually to fresh agar plates

Media composition. Medium A could be a standard growth medium (see Chapter 1) supplemented with additional organic components such as ribose and casein hydrolysate; auxin (e.g. 2,4 dichlorophenoxyacetic acid) at 1 mg/L and cytokinin (e.g. kinetin) at 0.1 mg/L. Osmoticum: sorbitol at 0.4 M. Medium B: standard medium with no added organics or osmoticum, auxin 0.1 mg/L; cytokinin 0.1 mg/L.

THE IMPORTANCE OF GENOTYPE

Some species yield protoplasts that grow and divide in culture very easily; protoplasts from some other species have resisted all attempts so far to induce even the first division. The reasons for this wide variation in response are unknown. It is an unfortunate fact that the species that are of most agricultural importance (e.g. cereals, legumes), and from which, therefore, we have most to gain through plant improvement programmes, have proved to be the least amenable to protoplast technology. One of the few generally applicable principles that has arisen from work on protoplasts is that of the importance of genotype in dictating ease or otherwise of protoplast isolation and culture. Even closely related cultivars can differ enormously in their response to protoplasting techniques. For example, in one study of the properties of cell cultures of over 60 varieties of maize, in only one case, Black Mexican Sweet, was it possible to culture the derived protoplasts and regenerate plantlets. The lesson arising from this type of study is that if the exact variety of a given species is relatively unimportant, as is often the case in more fundamental research projects, then time is better spent screening for amenable varieties, than in attempting to formulate conditions under which protoplasts from a given variety will undergo division and regeneration.

Biotechnological Manipulation of Protoplasts

The new era of plant biotechnology has been greatly facilitated by advances in culture-related technology that permit us to manipulate individual cells. Protoplast technology has given us access to the minimum viable unit of plant material, the isolated protoplast. The ability to manipulate cells and protoplasts has led to the development of three powerful techniques, each with the potential for facilitating the adaptation of crop species to current and forseeable agricultural needs. These three techniques are cell fusion, selection, and transformation. Only the first of these relates uniquely to protoplasts. In the other two cases the methods can apply to cultured cells (Chapter 5) and it must be ascertained in individual cases whether potential advantages in the use of protoplasts outweigh the considerable disadvantage of less predictable regeneration behaviour.

The major aim of these techniques is essentially the same; the restructuring of desirable plant characters into more favourable combinations. In addition, cell selection (in conjunction with chemical mutagenesis, or some other method of inducing genetic variation) and cell transformation have the potential for creating fundamentally new characteristics. The basic properties of protoplasts that make them attractive (or indispensable) for these applications are the absence of the rigid cell wall, a factor that allows cell–cell fusion and microinjection, and the fact that protoplasts exist as individuals as opposed to clusters in the case of cells, behaviour that greatly assists the production of true clones.

PROTOPLAST FUSION

In principle, protoplast fusion allows us to bring together any desirable plant traits (for example disease resistance, salt tolerance and high yield) in combinations that are not possible by sexual means. In other words, it is a technique that allows natural incompatibility barriers operating at the whole-plant level to be crossed. However, in reality, somewhat more subtle barriers to total genome integration have been encountered, especially during attempts to fuse distantly related parental protoplast types. These difficulties have rendered obsolete some early optimistic plant improvement schemes, but at the same time have uncovered some previously unsuspected lines of research. The true potential of protoplast fusion in crop improvement is still a matter of debate.

The Induction of Protoplast Fusion

Although fragile in comparison with an intact cell wall, the protoplast membrane still has a relatively high degree of structure stability. In order for protoplast-to-protoplast fusion to occur, protoplasts must be brought into intimate contact, and the membranes must be reversibly destabilized. Although some spontaneous fusion of protoplasts, as witnessed by the appearance of multicellular bodies, usually occurs during the isolation procedure, this process is not efficient or controllable enough for experimental use. Accordingly, specialized methods for protoplast fusion have been developed and of these chemical fusion, usually using poly(ethylene glycol), and electrically induced fusion (electrofusion) are the most important.

Before considering these methods in detail, we must define the terminology of protoplast fusion. Initially, fusion solely between protoplast membranes takes place and the resulting bodies are called *fusion products* or *heterokaryons*. If nuclear fusion takes place then the product is termed a *hybrid*. It is important to note that nuclear fusion does not always occur, but it is clearly the hybrid cells, i.e. those containing an integrated set of genetic material, with contributions from both parents, that are of interest.

Fusion is essentially a non-specific process. In general we cannot dictate which protoplasts will fuse. There are no constraints preventing self–self fusion (i.e. fusion of protoplasts from the same parent) or fusions between any number of individual protoplasts. However, the goal in protoplast fusion experiments is usually fusion between one protoplast of each parental type, so-called *binary fusions*, although at present we do not have techniques for guaranteeing this situation.

Chemically Induced Fusion

The most common fusing agent is poly(ethylene glycol) (PEG), although other substances (polyvinyl acetate, lipids) have been used in isolated instances. In fact certain protoplasts appear to be intrinsically very fusogenic and can be induced to fuse merely by exposure to high pH (10.5) and a high concentration of calcium ions. The reason for high intrinsic tendency to fuse is not known, but in the case of cultured cells it can be dependent on the conditions of culture. A more usual technique that is of wider applicability and less sensitive of variation is the use of high pH, high calcium conditions in conjunction with PEG.

At concentrations in the range of 10–50%, PEG causes extensive, non-specific agglutination of plant protoplasts and other cell types. Upon dilution with the appropriate washing solution (often a high pH, high calcium medium as described) a relatively high proportion of fusion (up to 100% has been claimed), can take place. Typically, however, fusion frequencies (the percentage of the total protoplasts present in fused units) lie in the range of 5–40%. The mechanism of action of PEG is not precisely known. The most plausible hypothesis is that the polymer, which is known to have a high affinity for water, immobilizes water molecules in the vicinity of the membranes, thus decreasing the hydrophobic stabilization of the lipid bilayers.

There are two main drawbacks associated with the use of PEG. Firstly, it is difficult to produce binary fusions in preference to multiple fusions, because of the tendency of PEG to induce large aggregates of agglutinated protoplasts. Secondly, PEG is fairly toxic and plating efficiencies following fusion can be low. Also regeneration frequencies, that is the number of protoplasts able to reform plantlets, can be lower still. Therefore, it appears that PEG can affect the pathways of regeneration in a non-lethal manner.

Electrofusion

The discovery in 1979 that protoplasts held in an appropriate electric field underwent fusion caused great excitement because it was immediately apparent that the process was much cleaner and more controllable than PEG-mediated fusion. Most major research groups working on protoplast fusion now use this technique, although PEG still finds an application when only a small number of fusions are necessary, because electrofusion machines are expensive.

As with PEG-induced fusion, the process consists of two phases, namely cell agglutination and membrane fusion (Fig. 3.4). To produce agglutination, a suspension of protoplasts is placed in a chamber between two metal electrodes. The chamber often takes the form of a modified microscope slide, so that the process can be followed visually, and the electrodes may be platinum wires perhaps 1 mm apart. When exposed to a non-uniform, alternating field, the protoplast membranes become differentially charged, and more positive regions are attracted to more negative regions on adjacent protoplasts. The membranes become charged in this way because the counter-ion cloud, associated with ionizable membrane groups, becomes distorted as a result of the imposed electric field. In consequence, protoplasts tend to line up in so-called pearl chains (Fig. 3.5), perpendicular to the electrodes. Once this protoplast–protoplast contact has been established, fusion is induced by a single high-voltage DC pulse. The mechanism underlying this effect is thought to be that the protoplast membrane, being an insulator, allows a large potential to build up between its inside and outside surfaces. If a protoplast is exposed to a sufficiently high voltage ($100\,kV/m^2$) then the induced charge difference between the inside and outside can be such that the force of attraction existing between the charged regions is sufficient to crush the membrane. On removal of the imposed potential, the original membrane structure is reformed, but in the case of agglutinated protoplasts, in which mixing of membrane components can occur in the destabilized state, formerly adjacent membranes can reform into a single continuous layer (Fig. 3.4), and the adjacent protoplasts therefore fuse.

Selection of Fusion Products
A typical fusion experiment will result in the products of binary fusions, self-to-self fusions, and multiple fusions, together with many unfused protoplasts and debris. The recognition and selection of the desired binary fusion products from this mixture constitutes a major problem in the application of protoplast technology. After an initial consideration of this problem, one might suppose that selection based on mutant complementation, such as has been supremely successful in the selection of monoclonal antibody-producing fused animal cells, might be the most appropriate technique. However, it turns out that for a variety of reasons, the most important of which is the genetic instability of plant cells in culture, stable mutant plant cells are extremely difficult to isolate and maintain. Although antimetabolite (e.g. 5-methyltryptophan)-resistant, auxotrophic, and conditional lethal mutant or variant plant cells have been obtained, their occurrence is not sufficiently widespread (Chapter 5) to make them suitable for general use.

The major techniques that have been used successfully in the isolation of protoplast fusion products are shown in Table 3.2. Of these techniques, manual isolation is by far the most widespread. In this technique it is required that the two parental protoplast types have distinct morphological markers, such as the presence of chloroplasts or starch grains, that are easily recognizable using a microscope. Fused cells bearing both markers can then be isolated using a Pasteur pipette, or an automatic pipette connected to a gentle suction. This method has the advantage that multiple fusion products can be rejected. The main disadvantage of the technique is that it is very time-consuming and tiring. A common way of obtaining different parental markers is to use mesophyll (green) protoplasts from one parent, and protoplasts from suspension cultures (colourless) of the other. The culture process may induce undesirable genetic variation in the latter parent, however. Starch grains, for use as a marker, can usually be induced in cultured cells by a period of growth on a high-sucrose medium during which the cells store excess carbohydrate. Pigmented protoplasts derived from petals have also found a use in manual selection techniques.

Notwithstanding the general limitations discussed earlier, mutant complementation-based selection has been used in some special cases. Possibly the most highly developed technique is the use of non-allelic albino mutants such that the fusion products or hybrids no longer show the albino trait and can be selected on

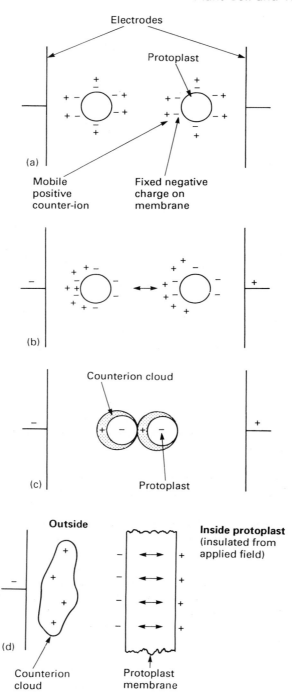

Electrodes

Protoplast

(a)

Mobile
positive
counter-ion

Fixed negative
charge on
membrane

(b)

Counterion cloud

(c)

Protoplast

Outside

Inside protoplast
(insulated from
applied field)

(d)

Counterion
cloud

Protoplast
membrane

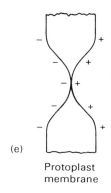

(e)

Protoplast
membrane

Collapsed area
of membranes

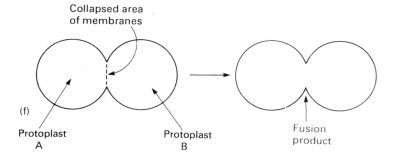

(f)

Protoplast
A

Protoplast
B

Fusion
product

Fig. 3.4 Electrofusion of protoplasts. The full mechanism of electrofusion of biological membranes is not known. The diagram shows a currently accepted model for the major events of the fusion process. (a) Protoplasts prior to application of electric field. (b) Under the influence of the field, the distribution of mobile counter ions to fixed membrane charges become distorted by attraction to the appropriate electrode. (c) The unequal distribution of charges around the protoplasts causes electrostatic attraction between the latter, resulting in pearl chain formation. (d) This diagram shows the induction of a potential across the protoplast membrane resulting from distortion of the counterions towards the electrode. Only one electrode is shown for clarity. The external face of the membrane becomes negative with respect to the inner face. Electrostatic attraction occurs between the oppositely charged faces as shown by the arrows. (e) When this force of attraction is sufficiently large the membrane collapses. (f) Areas of collapsed membrane between adhering protoplasts in pearl chains can reform into a continuous membrane, resulting in cell fusion, when the electric field is removed.

Table 3.2   The major methods for the selection of somatic fusion products
or hybrids

Selection method	Comments	Selects for[a]
Manual selection	Requires distinct morphological markers e.g. chloroplasts, starch grains or pigmentation	Fusion product
Mutant complementation	Mutants including those in nitrate reduction and chlorophyll synthesis pathways have been used	Hybrid
Altered morphology or growth requirements	Regenerating colonies may have a characteristic morphology or altered nutrient requirements	Hybrid
Fluorescence-activated cell sorting (FACS)	Rare and expensive	Fusion product
Altered physical properties	e.g. Buoyant density of fusion product may be different to parents	Fusion product
Antimetabolite resistance	Certain antibiotic and herbicide resistances are carried on the organelle genomes. Specific organelle combinations can therefore be selected for	Hybrid

[a] These assignments are not absolute. In many cases regenerating cell colonies consisting of a mixed population of the parental cell types can behave as a true hybrid culture.

(a)

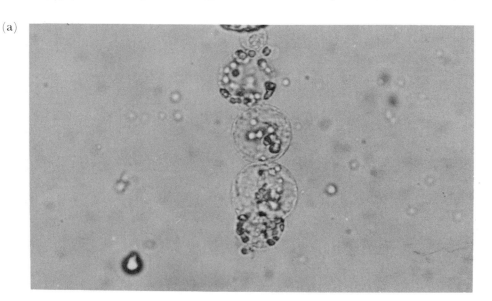

(b)

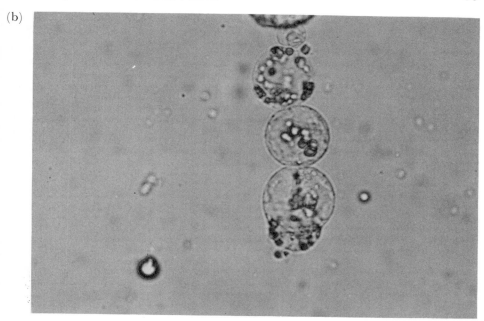

(c)

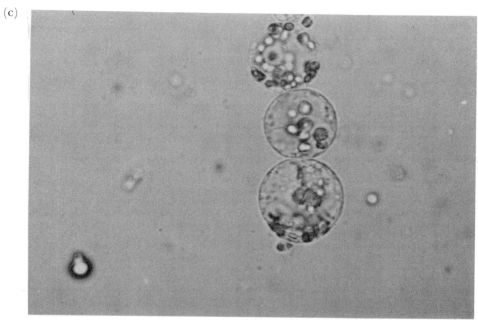

Fig. 3.5 Pearl chain formation and electrofusion of protoplasts. (a) 20 s after DC pulse, (b) 40 s after, (c) 20 min after. Photographs kindly supplied by C. Eady. The diameter of protoplasts is approximately 30 μm.

the basis of their greenness. Albino mutants are a favourable choice because they can be easily identified and isolated at the whole-plant level and do not require a selection step under potentially undesirable culture conditions. However, as with all mutant cells, the desirability of introducing potential metabolic defects into crop species has been questioned. Another point to bear in mind associated with the use of complementary mutants is that they do not necessarily behave the same when co-cultured (i.e. during selective conditions after fusion) as they do in monoculture. The possibility of cross-feeding effects, which could short-circuit the effect of the mutations, must be considered. As an example, selection based on auxin autotrophy can fail because auxin can diffuse from the autotrophic cells and supply auxin-requiring cells in the population.

A recent selection technique, originally developed for animal cell systems, is the use of fluorescently labelled cells in conjunction with a fluorescence activated cell sorter (FACS) (Fig. 3.6). The different parental protoplast types are labelled with different fluorochromes, commonly fluorescein and rhodamine, that fluoresce at different wavelengths. In the FACS, a given fluorescence activates a switch that directs the flow of liquid through the machine to one of a number of reservoirs. The dual fluorescence present in fusion products can thus be used to isolate them. One problem with this technique arises because adhering (not fused) protoplasts and multiple fusion products may all be recognized by the machine. Also, passage through the machine may cause extensive cellular damage on account of the fragility of protoplasts. This latter problem can be minimized if the protoplasts are first cultured for 24–48 h after fusion so that the regenerating wall affords some protection against physical damage in the FACS.

Many minor methods of fusion product selection have been applied, usually in particular cases in which special properties of the parental protoplasts lend themselves to the appropriate selection protocol. For example, hybrid vigour is occasionally expressed in fused protoplasts, and the products outgrow, and therefore can be selected from, the parental types. Similarly, in isolated cases hybrids have the ability to grow in the absence of auxin, a characteristic not expressed in the parents. In these cases fusion products can easily be isolated by growth on an auxin-free medium, providing cross-feeding can be minimized, for example by plating out at very low densities. Some physical selection procedures have also been developed for particular applications. As an example, sometimes the buoyant density of the fusion products is significantly different from the parental protoplasts, and this has allowed fusion product isolation by centrifugation in the appropriate medium.

Numerous other selection procedures have been reported, but none are suitable for general application. The development of an efficient selection method that is independent of the properties of the individual protoplasts, and that ideally discriminates against multiple fusion products would be a major advance in protoplast technology.

Proof of Hybridity
So far we have discussed mainly the immediate products of fusion experiments, that is those resulting from membrane fusions. However, these products only

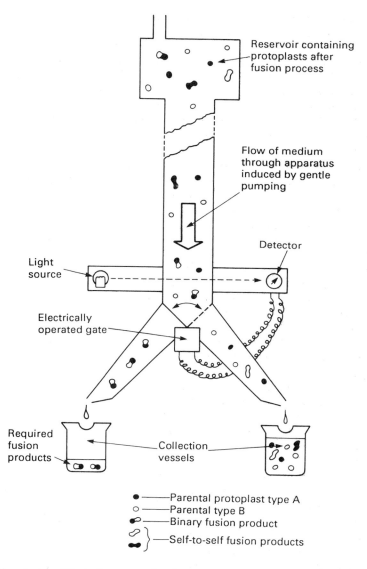

Reservoir containing protoplasts after fusion process

Flow of medium through apparatus induced by gentle pumping

Detector

Light source

Electrically operated gate

Required fusion products

Collection vessels

● ———— Parental protoplast type A
○ ———— Parental type B
● ———— Binary fusion product
} ———— Self-to-self fusion products

Fig. 3.6 A simplified diagram of a fluorescence-activated cell sorter (FACS).

become of interest to us if the subsequent step of nuclear fusion occurs, resulting in the formation of true somatic hybrids. For this reason, following fusion product selection we must also take steps to determine whether regenerant colonies or plants do in fact carry genetic information from both parents. In some cases the selection method used, for example mutant complementation, makes it highly likely that selected material is hybrid in nature.

The most straightforward method that suggests, but is not proof of, hybrid nature is the presence in selected plants of morphological characters that are intermediate with respect to the parents. Characters for which this type of behaviour has been observed include stalk height, leaf shape, and flower colour. The fact that these attributes show intermediate expression strongly suggests that they are polygenic traits. However, morphological characteristics are highly dependent on the growth conditions of the plants and so careful experimentation is required. Furthermore, such traits may vary due to culture-induced variation in unfused protoplast-derived plants. Intermediate morphology is therefore useful only as a first indication of hybridity.

If the chromosomes of the two parental genotypes differ markedly in size or morphology then it may be possible to identify hybrids by karyotype analysis. However, such clear-cut difference in chromosome structure are relatively rare.

By far the most common method for the confirmation of hybrid status is isoenzyme analysis. A given enzyme usually has a unique, species-specific amino acid sequence. The differences between species are often very minor, but can be sufficient to alter the behaviour of the protein during electrophoresis, such that distinctive bands characteristic of species origin can be detected by activity staining. Enzyme bands rather than total protein profiles are used because the latter are far too complicated to allow easy interpretation. Enzymes commonly used include esterases, peroxidase, and alcohol dehydrogenase. However, care must be exercised in the interpretation of isoenzyme patterns because they can vary quite markedly between the different tissues in a given species, and in the same tissue at different development states. Sometimes isoenzyme bands that do not correspond to the parental bands are seen on electrophoretograms of samples from hybrid tissue. This behaviour has been observed with aspartate aminotransferase, a dimeric enzyme, and has been interpreted as resulting from a hybrid dimer containing one subunit from each parental cell type.

Secondary metabolites are often characteristic of species and indeed are used in taxonomic classification. Accordingly, secondary metabolite profiles have been used in a similar way to isoenzyme analysis for the confirmation of hybrid status, but only in isolated instances as yet.

In principle, the most convincing test for parentage of a supposed hybrid material would be DNA restriction enzyme fingerprinting, as recently applied to human DNA. Techniques have only recently been developed that can accommodate the large size of the plant genome, and these have not yet been applied to the analysis of protoplast fusion products. However, the smaller chloroplast and mitochondrial genomes have been amenable to this approach for several years, and it has proved possible to determine the origin of cytoplasmic hybrids (see below) by this method.

Genetic Consequences of Protoplast Fusion

Protoplast fusion at the level of the membrane is a non-specific process, at least at the present level of understanding; consequently protoplasts from any plant species (and even animal and plant) can be fused. Likewise, nuclear fusion does not appear to depend upon any special characteristics of the parental cells. When we consider events inside the hybrid nucleus, however, the situation changes and it is clear that certain incompatibility reactions (somatic incompatibility) can exist. These reactions are likely to occur as a consequence of the normal ontogeny of the nucleus and not by the operation of specific incompatibility mechanisms, because cell fusion rarely occurs in plants in nature.

The most obvious result of protoplast fusion is an additive increase in the chromosome complement of the hybrid cell. This situation rarely, if ever, persists when the protoplast is cultured, and when a cell colony or eventually a plant is derived, it is usually found that some or most chromosomes from one parent are preferentially lost. This loss frequently becomes apparent as a progressive loss of enzyme bands from electrophoretograms. The mechanisms governing which set of chromosomes is to be partially eliminated are still not well understood, but recently two general principles have emerged. The more obvious of these principles is the concept of genome competition. It has been found that the chromosomes of the parent with the shorter cell cycle time are often retained. This suggests selection on the basis of direct competition. We can infer from this that any factors conferring a selective advantage on a given chromosome or set of chromosomes may lead to the retention of that genetic material. The second general principle is that of compatible (or incompatible) states of the parental genomes. It has been observed that the more dissimilar the state of the genomes, the greater is the extent of chromosome elimination. For example, when protoplasts at different phases of the division cycle are fused, the chromosomes can be at different stages of condensation. In this situation premature condensation of the less condensed chromosome set has been observed, and this type of overriding of the nuclear replication programme may result in a perturbation of the normal controls governing chromosome duplication. A second possible case of genome incompatibility exists when cells of different states of differentiation are fused. Within a plant, different cell types have different portions of their genomes expressed, the remainder being 'turned off', for example by methylation. The presence in a single (fused) genome of homologous chromosome regions both turned on and off may create a conflict that results in upset to the mitotic cycle, resulting in chromosome loss. It must be stressed that hard evidence for the above mechanisms is scant, but these hypotheses at least provide a conceptual framework for viewing potential causes of genome instability.

Somatic hybrids that lose chromosomes from one parent, and therefore are phenotypically closer to the other parent are called asymmetric. The rarer hybrids that retain substantial amounts of genetic material from both parents are termed symmetric. Perhaps the most widely publicized somatic hybrid is 'Arabidobrassica', a cross between *Arabidopsis thaliana* (10 chromosomes) and *Brassica campestris* (20 chromosomes), both of the crucifer family. The hybridization of these species resulted in the formation of a range of products with chromosome numbers from 35 to 80, and in this case the hybrids could be asymmetric with respect to either parent.

Chromosome elimination is not restricted to fusion hybrids of relatively distantly related species, however. Certain sexual hybrids exhibit a similar behaviour. It is significant here that in cases in which the equivalent somatic hybrids have been produced by protoplast fusion, parallel chromosome loss has been observed. This finding suggests that in some cases at least, chromosome loss is not a haphazard process, and could even be predicted if we had a sufficient level of understanding of the mechanisms involved.

Following the progressive loss of certain chromosomes from a fusion hybrid, a (relatively) genetically stable state is usually reached. From a biotechnological point of view this is in fact a very favourable situation. Using protoplast fusion we cannot hope to create a 'super plant' containing all the desired characteristics at a single step. Some properties of the parents that were already optimized would almost certainly be impaired by the fusion process, e.g. by recombination. Rather, the aim is to produce in an essentially unchanged parent a single enhanced trait, or a very small number of them. Chromosome elimination is a process that tends towards this very situation, in which basically one complete parental genome is retained, with the addition of only a small number of elements from the second parent. The present uncontrollable and unpredictable nature of chromosome loss means that a considerable amount of selection for the required genotype must follow the recovery of hybrid material.

So far we have considered the genetic behaviour of hybrids at the purely chromosomal level, but the actual situation is certainly far more complex than this. There is evidence that in a somatic hybrid, as in a sexual hybrid, intergenomic recombination, i.e. the rearrangement of genetic material between chromosomes of the two parents, also occurs. The extent to which this happens, and the nature of the associated controlling factors are unknown, but recombination is a potential source of phenotypic variation because the expression of a gene is known to respond to the genetic environment in which the gene is located.

Chromosome Elimination and Somatic Cell Genetics

The phenomenon of chromosome loss from fusion hybrids, whilst introducing an element of uncertainty into biotechnological studies, in fact proves to be of the utmost usefulness in fundamental investigations into the coding functions of individual chromosomes. By correlating chromosome loss with the disappearance of phenotypic characters (e.g. morphological traits, metabolic activities, or individual enzymes) from the cells or regenerant plants, researchers have been able to assign the characters to specific chromosomes. This technique, called somatic cell genetics, has also proved extremely useful in probing the human genome.

The Cytoplasmic Genomes

A further level of complexity relating to the genetic behaviour of fusion hybrids is revealed when we consider the fate of the cytoplasmic genomes, for our purposes those of the mitochrondria and chloroplasts, following the fusion events. Traits such as some types of male sterility, disease resistance, and herbicide resistance are known to be coded on cytoplasmic genes. Male sterility is important because it effectively prevents unwanted self-fertilization, and this facilitates desired out-crossings which can be important for the efficient running of a breeding programme. For reasons that are not fully understood, the different parental types

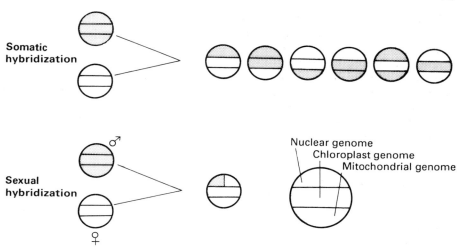

Fig. 3.7 A diagrammatic illustration of some novel genome combinations made possible by somatic hybridization. The nuclear genome of one somatic cell type of the fusion pair can be inactivated by X-irradiation, the cytoplasmic DNA being more resistant to damage. Additional variation is available due to incomplete segregation of organelles, resulting in the presence of organelles from both parents, and also through cytoplasmic genome recombination. In sexual hybridization, the inheritance of the cytoplasmic genomes is thought to be predominantly maternal.

of chloroplasts and mitochondria do not become fully mixed following fusion, rather they segregate and compete, such that hybrids containing exclusively one parental type or the other are produced. Furthermore, chloroplasts and mitochondria assort independently and consequently a much wider range of nucleus–cytoplasm combinations arise than are seen in sexual crosses in which uniparental contribution of organelles is often observed. This process creates diversity out of which potentially useful gene combinations may arise. Superimposed upon this picture is the phenomenon of genome recombination between organelles. This has been proven to occur between mitochondria, but there is little evidence as yet that the same process occurs between chloroplasts. Clearly cytoplasmic genome recombination greatly enhances the potential genetic diversity that can be achieved and explored through the use of protoplast fusion technology.

Cytoplasmic genomes can be manipulated in a unique fashion by the technique of cybridization. In this variant of protoplast fusion, the nuclear genome of one parental protoplast type is inactivated, for example by X-rays. The subsequent induction of fusion, coupled with the associated organelle segregation, produces a range of cytoplasmic variants around the theme of the intact parental nucleus. In this way cytoplasmically encoded traits can be manipulated specifically, whilst leaving the majority of the plant's characters (encoded by the nucleus) intact. Many researchers see more potential in this technique than in standard protoplast fusion itself.

Analysis of the isoenzymes of ribulose bisphosphate carboxylase, a chloroplastic enzyme functioning in photosynthetic carbon fixation, has been especially useful in the characterization of cytoplasmic hybrids. The enzyme is composed of two types of subunits, the small subunit being encoded by nuclear genes whereas the large unit is chloroplast-specified. Isoenzyme analysis therefore reveals the parentage of the hybrid cells.

The Biotechnological Potential of Protoplast Fusion
Few biotechnologists would disagree with the view that protoplast fusion technology has potential for the improvement of commercially important plant species. However, we need to assess the significance of protoplast fusion in the knowledge that, despite intense effort, there have been few commercially relevant products of these techniques. We also need to realize that traditional plant breeding techniques are very successful and will not easily be supplanted. Rather, the contribution that protoplast fusion techniques may make to agriculture will complement and not replace these existing procedures.

The major limitations preventing the widespread application of protoplast fusion technology are the difficulty in regeneration of many fusion products, and the often low fertility of plants that have been recovered. The plant regeneration problem is not uniquely a consequence of the fusion process (but may be compounded by it), but is to some extent inherent in the isolated protoplasts themselves, although there is no reason to believe that these difficulties cannot be overcome with the appropriate refinements in methodology. In fact substantial improvements have been made in the last decade, and perhaps 50–60 intraspecific, interspecific, and even intergeneric fusion hybrids have now been regenerated into plants. The crop species that can be regenerated readily from protoplasts include alfalfa, asparagus, cabbage, carrot, cassava, citrus, clover, endive, millet, potato, and sunflower. Unfortunately, the species that constitute the most important crops, for example the legumes and cereals, are extremely difficult to regenerate. The mechanisms controlling cell division in protoplasts and the subsequent onset of organized growth and plant regeneration are still obscure, and at the moment the prospect of the rational design of protocols (as opposed to an empirical approach) to enhance the recovery of plant material seems remote.

To consider the potential of protoplast fusion in agriculture let us return to the question of what it is that fusion actually achieves, and equate this with the needs of a plant improvement programme. It is important to realize that crop improvement is a commercial enterprise and in a given situation the technique used will be that which produces the end result most cheaply, rather than that which is most scientifically satisfying or technically clever. Therefore the possible gains due to fusion of protoplasts must be viewed alongside other ways of achieving the same results. We have said that protoplast fusion can bring together desirable plant traits in combinations that are not possible by sexual means, i.e. sexual incompatibility barriers can be crossed. However, certain types of incompatibility

Table 3.3 Improvement of crop species using protoplast fusion. These hybrids are not of direct agricultural importance but are being evaluated as sources of desirable traits for further breeding programmes

Character	Source	Regenerated hybrid
Atrazine resistance Cytoplasmic male sterility	Different *Brassica napus* cultivars	*B. napus* × *B. napus* (rape)
Atrazine resistance	*Solanum nigrum*	*S. nigrum* × *Lycopersicon esculentum* (tomato)
Ability to regenerate from protoplasts	*L. esculentum* (wild)	*L. esculentum* (wild) × *L. esculentum* (cultivated)
Disease resistance	*Glycine canescens*	*G. canescens* × *G. max* (soybean)
Virus resistance	*Solanum brevidens*	*S. brevidens* × *S. tuberosum* (potato)
Cytoplasmic male sterility	*Beta maritima*	*B. maritima* × *B. vulgaris* (beetroot)

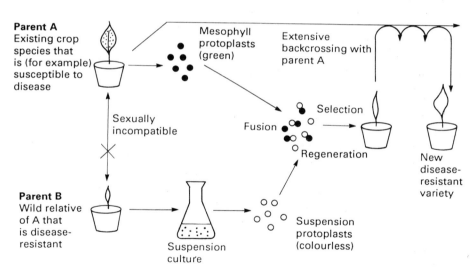

Fig. 3.8 A strategy for the production of an improved crop variety.

can be overcome by other means. For example, some potentially desirable sexual crosses produce embryos that abort before mature fertile seed can be produced. It is possible in some cases to rescue these embryos by excision and subsequent culture under a rich nutrient regime. This procedure often works when abortion is due to failure of the endosperm to provide all the nutrients required for normal growth and development.

Although in principle any characters can be combined by protoplast fusion, we have seen that improvement in one character is unlikely to be achieved, as the result of a single fusion, without the downgrading of other characters previously optimized. This is because of the gross genetic upset that whole genome hybridization produces. Workers in the field of somatic hybridization are aiming for the creation of novel (not necessarily optimized) combinations of characters that can subsequently be entered into a conventional breeding programme. The desirable characteristics can then be 'quietly' expressed in an already optimized genetic environment by backcrossing with the appropriate parent. When compatibility with a breeding scheme is recognized as a vital attribute of a somatic hybrid, some constraints on the latter are imposed. The pollen fertility of somatic hybrids is often low. A more or less direct relationship has been found between the pollen viability and the relatedness of the parental species. Very low pollen viability is usually found in interspecific hybrids, and intergeneric hybrids are usually sterile. An additional complication has been encountered in that the ability of viable somatic hybrid pollen to cross can be restricted. For example, hybrid pollens have been recovered that do not permit selfing and others that do not permit backcrossing. The low viability and lack of flexibility with respect to crossing generally make those hybrids expressing these properties unsuitable for conventional breeding programmes.

In certain instances low pollen fertility is not a barrier to the biotechnological exploitation of hybrid species. In vegetatively propagated plants, clearly, these sex-related factors do not apply. However, in this case the desired genotype must arise directly from the fusion procedure. A second instance is that of the production of commercially useful cell lines in culture. As discussed in Chapter 6 many cell lines can produce useful compounds, e.g. pharmaceuticals, food-related products, etc., and these substances can, in principle at least, be produced on a large scale in bioreactors (Chapter 9). The manipulation of cell lines to synthesize larger amounts of these products, or to combine production with other characteristics such as high growth rate, ability to excrete product, etc., is perhaps the most immediate application of fusion technology. This is because the fertility problem, and indeed regeneration itself is not involved. In contrast to the situation with respect to intergeneric hybrid plants, many stable intergeneric hybrid cell lines have been produced.

In one interesting case protoplasts of two species, one (*Euphorbia millii*) producing the food colour anthocyanin and the other (*Coptis japonica*) producing the pharmaceutical berberine were fused, resulting in a hybrid cell line that produced both substances. Use of this type of somatic hybrid could result in savings in the cost of production in bioreactors by essentially sharing the cost over two or more products.

Wild relatives of cultivated crops are often relatively resistant to diseases and pests. For example, the wild relative (*Solanum brevidens*) of potato is a source of virus resistance. The very fact of their relatedness makes the possibility of transfer of resistance to susceptible cultivated varieties by somatic hybridization an attractive proposition because genome disruption including subsequent fertility effects may be small. However, quite large ploidy differences can exist between wild and cultivated varieties and this can cause genome incompatibility problems.

Protoplast fusion is only one of the emerging techniques that is causing excitement because of the potential for enhancing the power of plant improvement programmes. The engineering of plant traits by specific gene manipulation is perhaps the other main procedure (Chapter 5). What is the relationship between fusion and gene manipulation methodologies, and which technique is preferable? Protoplast fusion is concerned with the gross hybridization of genomes and the subsequent selection of useful gene combinations, and it is largely a matter of chance if the correct combination of traits turns up. The power of protoplast fusion is that we do not need to understand the mechanism underlying the characteristics we are dealing with. Specific gene manipulation, on the other hand, allows the addition of one gene, or a small number of genes, to an already near-optimum species with only a low risk of impairing the existing characteristics. The drawbacks associated with gene transfer are that some understanding of the relevant traits is usually necessary, and also that polygenic characters may be extremely difficult to manipulate in this way. Consider, for example, the transfer of a biosynthetic pathway to a deficient species. Many enzymes will be involved, and the corresponding genes may be scattered throughout a chromosome or even across several chromosomes. These genes may be expressed to different levels, possibly at different times and probably in different cellular compartments. We simply do not have sufficient sophistication of either knowledge or technique to achieve this level of control of artificially inserted genes. In contrast, it has been possible to transfer biosynthetic pathways using protoplast fusion although we have little control of the overall constitution of the resulting hybrid.

In conclusion, protoplast fusion and gene manipulation are at present complementary skills, but in both cases their effective implementation awaits further advances in our understanding both of the techniques themselves and the characteristics we wish to manipulate.

OTHER BIOTECHNOLOGICAL APPLICATIONS OF PROTOPLASTS

As we have seen, the characteristics of protoplasts that make them attractive for a variety of manipulations are the lack of the physical barrier imposed by the wall, and the fact that protoplasts exist as individual units, not as multicellular aggregates as found in virtually all cell cultures.

In addition to permitting fusion, the former attribute allows the delivery into protoplasts of substances that cannot enter cells because of the relatively low porosity of the wall. The disaggregated state of protoplast preparations allows much more stringent selection procedures to be imposed, because of the lessened risk of the metabolic cooperation that occurs between cells in aggregates. The biotechnological operations that these properties facilitate are genetic transformation and cell selection.

Genetic Transformation Using Protoplasts
The addition of specific genes to a plant species in order to enhance or augment its properties is a technique with almost limitless potential. The same technique applied to the study of the basic cell biology of plants likewise gives an approach of the utmost power. The applications of gene manipulation techniques to plant

improvement are considered in Chapter 5, and do not necessarily require the use of protoplasts. For example, gene vectors based on plasmids of *Agrobacterium* species or plant viruses can also be applied to cell cultures and whole plant material, and these last systems are usually preferred because regeneration of whole transformed plants is likely to be far easier than from protoplasts. However, certain physical (as opposed to biological) transformation techniques, namely electroporation and microinjection, do need protoplasts for optimum efficiency.

Electroporation
Electroporation is basically a variant of electrofusion in which the electrical parameters, including field strength and pulse duration are adjusted such that short-lived pores are formed in the protoplast membrane, allowing the entry of molecules such as DNA that are normally excluded. Precise control of the duration of pore opening is required so that the molecule to be delivered can enter in sufficient quantities, but the viability of the protoplast is not irreversibly impaired. It has been found by experiments with dyes that pores can take up to 30 min to reseal completely, but this surprisingly does not kill the protoplast.

Unlike the use of *Agrobacterium* plasmid-based vectors, electroporation does not necessarily result in the added DNA entering the nucleus and becoming incorporated into the host DNA. *Agrobacterium* T-DNA naturally has a mechanism for achieving this. Non-incorporation has the disadvantage that the encoded genes may not be stably expressed or inherited. However, even in cases in which stable incorporation does not occur, electrically introduced genetic information is frequently expressed at least for a limited period in the electroporated cells, and this 'transient expression' has become a very useful rapid technique for exploring the sequence requirements for gene expression in plants.

DNA can also be induced to enter protoplasts by treatment with chemicals such as PEG although these methods are not as efficient as electroporation. The mechanistic details of these processes are not fully understood, but almost certainly charge neutralization of nucleic acid and protoplast membrane (both negatively charged) plays some part.

Electroporation has also been used to deliver other molecules, such as metabolic intermediates and viruses, to protoplasts. In the former case, electroporation allows the exploration of metabolic pathways by the addition of radioactively labelled substances that cannot normally cross the cellular membrane.

Microinjection using an appropriate syringe is another method for the direct addition of molecules including DNA into plant cells. Although protoplasts are not absolutely essential in this technique, their use facilitates the injection process because the wall of intact cells readily blunts or deflects microinjection needles. Given sufficient manual skill, material can be injected directly into the nucleus, which might enhance integration rates.

The Use of Protoplasts in the Selection of Desirable Genotypes
This topic will be considered in detail in Chapter 5. The techniques apply almost equally well to cells or protoplasts, but the latter have the advantage that it is easier to be confident that selected colonies are of single cell origin, i.e. are true

clones. Plantlets regenerated from protoplasts of either whole plant or culture origin have exhibited variation in a wide range of agriculturally important characteristics including disease resistance, yield, salt tolerance, etc. In some cases these properties have been enhanced with respect to the parent plant, indicating a possible means of crop improvement as will be discussed in Chapter 5.

References and Further Reading

Alfonso, C.L., Hawkins, K.R., Thomas-Compton, M.A. *et al.* (1985). Selection of somatic hybrid plants in *Nicotiana* through fluorescence-activated sorting of protoplasts. *Biotechnology* **3**, 811–816.

Evans, D.A. (1983a). Protoplast function. In *Handbook of Plant Cell Culture* (D.A. Evans, W.R. Sharp, P. V. Ammirato, and Y. Yamada, eds). Macmillan, New York. (Contains many interesting articles.)

Evans, D.A. (1983b). Agricultural applications of plant protoplast fusion. *Biotechnology* **1**, 253–261.

Fowke, L.C., and Gamborg, O.L. (1980). Applications of protoplasts to the study of plant cells. *International Review of Cytology* **68**, 9–51.

Galun, E. (1981). Plant protoplasts as physiological tools. *Annual Review of Plant Physiology* **32**, 237–266.

Gleba, Y.Y. and Hoffman, F. (1980). 'Arabidobrassica': a novel plant obtained by protoplast fusion. *Planta* **149**, 112–117.

Morikawa, H. and Yamada, Y. (1985). Capillary microinjection into protoplasts and intranuclear localisation of injected materials. *Plant Cell Physiology* **26**, 229–236.

Potrykus, I., Harms, C.T., Hinnen, A. *et al.* (eds) (1983). *Protoplasts 1983.* Proceedings of the 6th International Protoplast Symposium, Basel. Birkhauser, Basel. (Contains many useful articles, especially those by M.R. Davey and G.B. Lazar.)

Zimmerman, U. and Scheurich, P. (1981). High frequency fusion of plant protoplasts by electric fields. *Planta* **151**, 26–32.

Chapter 4

The Regeneration of Plants from Cultured Cells and Tissues

GRAHAM WARREN

Introduction

The ability to regenerate whole, fertile plants from selected or engineered cells is the key that unlocks the vast biotechnological potential of plant cell culture.

The induction of plant tissues to reform whole plants is of course not a new process but has been recognized and exploited by gardeners and horticulturalists for a very long time. However, over 40 years of research of a more formal nature, at an increasing degree of sophistication, have led us to the present stage where virtually any plant species can be regenerated, at least from isolated tissues, and scientists are beginning to contemplate seriously such concepts as artificial seed production, and mass, mechanized embryogenesis.

But these are certainly not grounds for complacency. Plant regeneration from levels below that of the isolated tissue, namely from cultured cells and protoplasts, is far less certain, and for some of the most important crops, for instance legumes and cereals, is rarely possible. This situation reflects the lack of basic understanding of plant differentiation and development, and of the control of growth in general. The mechanism of action of plant growth regulators (hormones) has been exceptionally difficult to study, and there is very little real prospect of this situation changing radically in the near future. None the less, progress is slowly being made, usually by following an empirical experimental approach, and there appears to be no theoretical reason why in the future any species may not be regenerated from isolated cells and protoplasts with relative ease.

Returning to the present, the major uses of plant regeneration are in plant propagation, the recovery of plants from engineered or selected cells, the production of disease-free stock, germplasm storage, and fundamental investigations into plant developmental biology.

Differentiation in Cultured Plant Cells and Tissues

One of the great unsolved problems in biology is the process by which single cells (e.g. germ cells), or small populations of seemingly identical cells undergo the co-ordinated division and development that result in the formation of a complex, highly structured mature organism in which a great many different cell types may be present. This is the process of differentiation, and clearly involves the differential expression of genetic information. However, the mechanisms by which individual genes are turned on and off, and especially the factors governing the temporal scheme of gene activation and deactivation, have proved elusive. Much more is known about differentiation in animal and micro-organism systems than in plants, mainly because of the better experimental systems available, but even here the level of knowledge is low. We therefore have to view differentiation in cultured plant cells and tissues from a standpoint of relative ignorance.

THE TYPES OF DIFFERENTIATION SEEN IN CULTURE

The main types of differentiation seen in plant cell and tissue cultures are root formation and shoot production, collectively termed *organogenesis*. In a smaller number of cases these two processes occur simultaneously in an apparently co-ordinated manner (as in the plant embryo) and this phenomenon is called somatic or adventive embryogenesis. In a few instances somatic embryos follow a sequence of development that is strikingly similar to that of a zygotic embryo, but in other cases the distinction between true embryogenesis and localized root and shoot formation can be rather blurred. The formation of roots, shoots, and embryos is of course a gross form of differentiation that encompasses a large number of individual biochemical events. The fact that intermediate morphogenetic pathways do not often occur suggests that the mechanisms co-ordinating differentiation are subject to tight internal control, but that the switches that direct the cells in question along a given developmental pathway are accessible to manipulation. These types of differentiation are the main morphological changes seen in cultures, but it is likely that a variety of biochemical differentiation pathways, not leading to morphologically distinct products, also occur. Some other types of morphological differentiation are occasionally seen, of which the formation of vascular tissue is perhaps the most common, but this can be associated with organogenesis.

Both organogenesis and somatic embryogenesis can result in plantlet formation. In the former case, shoot formation usually precedes root formation, and vascular differentiation is subsequently induced in the intervening cells such that eventually a root–shoot connection is made, and an autonomous plantlet is created. Experiments on the control of vascularization have shown that the linking up of root and shoot tissues is probably co-ordinated by diffusible hormonal signals released from the respective meristems.

THE ORIGIN OF ORGANOGENESIS AND EMBRYOGENESIS IN CULTURE

On account of the fact that cell cultures usually exist in an aggregated state, and because they are relatively variable with respect to cell size and shape, it has proved difficult to define the sequence of events leading to organ or embryo formation. For this reason there is still controversy over whether these structures arise from single cells or cell groups. It is similarly uncertain whether all morphogenetically competent cells in cultures are direct descendants of such cells present in the original explant, or whether offspring of any explant cell can give rise to differentiated structures. In a very small number of cases embryo development has been traced from a single cell, but this does not prove the generality of the mechanism. Indeed in a few cases chimeric plants have been obtained from mixed cell suspensions via embryogenesis, behaviour that suggests a multicellular origin of the somatic embryo. Similar observations have not been made during organogenesis.

In the case of tissue (as opposed to cell) culture the meristematic regions that give rise to morphogenesis arise either by multiplication of existing meristems (axillary meristems) or are produced *de novo* from other competent cells in the tissue by the formation of adventitious meristems.

EMBRYOGENESIS

The sequence of cell divisions and developmental events that gives rise to a somatic embryo in culture usually shows far more variation than the equivalent process in the ovule. Probably the main reason for this is the lack of whole-plant control systems in the culture environment, and the result is that somatic embryos exhibit a much greater range of sizes and shapes than zygotic embryos. Stages in embryogenesis of cultured carrot cells are shown in Fig. 4.1. This system is one of the most controllable somatic embryogenesis processes, and is often used as a model system for studies into cell differentiation. Species from many genera will undergo the early stages of embryogenesis up to the torpedo stage, but fewer (at present) can be induced to develop on into plantlets.

Embryogenesis has been formally defined as the formation of a structure in which a single shoot and a single root pole develop in the same temporal sequence as that seen in the zygotic embryo. In practice this definition is virtually impossible to apply fully, and normally any structure in which a root and shoot develop more or less synchronously is taken to be an embryo.

Somatic embryogenesis (Fig. 4.2) is not merely an artefact of the culture process; it also occurs in nature, for example from nucellar tissue in *Citrus*. This example also provides an illustration of the effect of genotype on somatic embryo formation. Species of *Citrus* that are naturally polyembryonic produce cultures that are highly embryogenic, whereas species that are monoembryonic usually produce few somatic embryos.

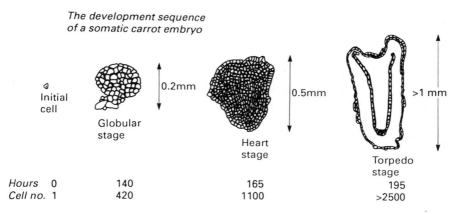

The development sequence
of a somatic carrot embryo

Initial cell	Globular stage	Heart stage	Torpedo stage
Hours 0	140	165	195
Cell no. 1	420	1100	>2500

0.2mm   0.5mm   >1 mm

Fig. 4.1 The sequence of development of somatic embryos in suspension cultures of carrot cells. This series of longitudinal sections represents development that is considered to be typical. However, in suspension cultures many misshapen and multiple embryos are also seen.

The usual trigger for embryogenesis in cultures is the removal of auxin, or the substitution of a less potent auxin, e.g. NAA, for a more potent one, eg. 2,4-D. Many cultures need a high-auxin (usually 2,4-D) treatment prior to this triggering step in order to achieve the rapid rate of cell division (meristem-like conditions) required for embryo initiation. The latter stages of embryogenesis are usually hormone-autonomous (i.e. produce sufficient hormones endogenously). In fact the addition of hormones at this time usually disrupts embryo development.

Two distinct types of somatic embryogenesis have been recognized. Direct embryogenesis, in which a single cell (or cell group) commences meristematic growth and all the descendants of this cell form part of the embryo, is a rare event. It has been reported to occur in *Citrus* nucellar tissue, in some anther cultures and very occasionally from protoplasts. Much more common is indirect embryogenesis, in which an embryo develops from one cell in a previously formed meristematic cluster (Fig. 4.3). Typically in such clusters only a few surface cells give rise to embryos, although it is thought that all the cells in the cluster are competent to undergo embryogenesis. This raises the possibility that some inhibitory response, akin to apical dominance, is initiated at an early stage in embryo development.

Several research groups have investigated the possibility of using plant somatic embryogenesis for the study of cell differentiation and development. In principle, plant embryo systems have several advantages when compared with the commonly used animal systems, for example amphibian embryos. The pattern of cells in plant embryos is far simpler than that in a typical animal embryo and there are far fewer different cell types. Furthermore, the cell migration observed during amphibian embryo development, a phenomenon that obscures the relationship between cell position and developmental fate, is not seen in plants. However,

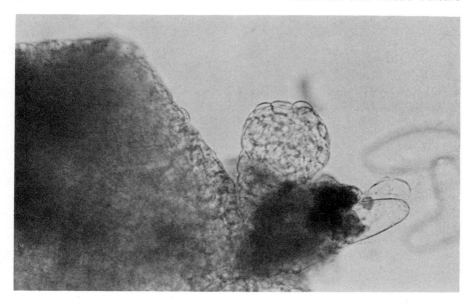

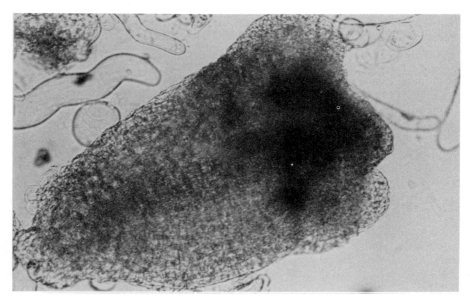

Fig. 4.2   Globular and torpedo-stage somatic embryos of carrot.

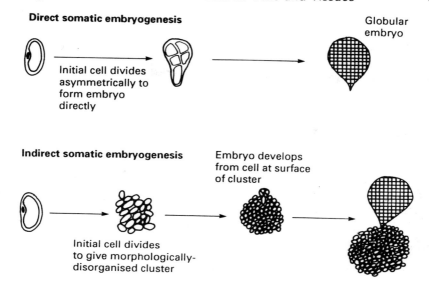

Fig. 4.3 Direct and indirect somatic embryogenesis.

despite these advantages, little progress has been made towards the understanding of cell differentiation through the use of plant embryo systems. Two factors are largely responsible for this. First, the sequence of cell divisions during somatic embryogenesis in plants tends to be highly variable and there are therefore problems in obtaining uniform experimental material. Second, populations of pre-embryonic cells, which would provide an extremely useful comparison with cells committed to embryogenesis, are not available because only a relatively small number of cells undergo embryogenesis, and it is not possible to identify and isolate these cells prior to embryo formation.

Somatic embryogenesis in carrot cultures featured as one of the scientific experiments performed during a joint USA/USSR space mission. Successful embryo development suggested that gravitational polarization is not a requirement for this process.

ORGANOGENESIS

Organogenesis, for our purposes root or shoot formation, is a more widespread and controllable process than embryogenesis. In certain instances, e.g. tobacco, fairly precise hormonal switches are available that determine the developmental pathway followed, making these useful systems for study of the control mechanisms involved. It may be the case that root and shoot formation can be considered as facets of the same process because it has been reported (e.g. with *Convolulus*) that the same meristem or primordium can be induced to form a root or a shoot depending on the growth conditions.

As is the case with other types of differentiation, the main control we have over organogenesis is by application of the appropriate hormones. Although the exact nature of these hormonal triggers can vary greatly between species, the ratio of auxin to cytokinin has been found to have a consistent effect in a variety of systems. Hormonal effects on differentiation will be considered on pp. 90–91.

Organogenesis from cultured cells depends on the presence of pre-existing (i.e. carried through from the explant) or induced meristematic primordia. These primordia are rapidly dividing groups of cells in which there is presumably a degree of spatial/biochemical organization conducive to differentiation. Meristematic cell clusters can arise from relatively dormant vacuolated cultured cells under the appropriate conditions, usually the presence of high auxin levels. However, it is more usual for meristematic and vacuolated cells to behave as independent populations.

In the case of cultured tissues, pre-existing axillary meristems can be induced to proliferate by the removal of apical dominance effects. This is usually achieved by the application of cytokinins. Adventitious meristems can be induced from explant cells either directly, or with an intervening callus stage. The pathway followed can have a dramatic effect on the genetic constitution of the regenerated plantlets (pp. 92 and 98).

HORMONAL EFFECTS ON DEVELOPMENT

Plant growth regulators do not always behave as typical hormones (e.g. they can exert effects in the cell in which they are produced) and for this reason many workers are reluctant to apply the term 'hormone' to these substances. However, these molecules do also exhibit true hormonal effects (action at a distance) and so the term 'plant hormone' will be retained here for convenience.

Plant hormones have such profound effects on cell differentiation and development that for a long time many researchers believed that these molecules were the primary agents responsible for differential gene expression during morphogenesis. There is now a growing feeling that this is unlikely to be the case, first because hormones appear able only to switch between 'preset' patterns of morphological development, a situation suggesting that other classes of molecule are involved in primary differentiation events, and second because hormones can have very different effects on a given tissue in different species, or on the same tissue at different developmental states. There is now a view that plant hormones constitute a pool of intra- and intercellular messengers that can be utilized according to the demands of the cell or tissue at a particular stage, and which therefore, exhibit a changing spectrum of partially overlapping functions. In view of this situation, it is rarely possible to extract general principles underlying hormone action or, as yet, to identify the primary molecular mechanisms involved.

There is a vast literature concerning the effects of plant hormones on whole-plant tissues and cell cultures, and many of the experimental results have been contradictory. Recent advances in understanding of cell–hormone interaction have led to the recognition that there is unlikely to be a simple relationship between the concentration of a hormone added to a tissue or cell, and the actual effective concentration at the site of action.

Table 4.1 Some plant growth regulators (hormones) that are commonly added to plant cell cultures

Class	Growth regulator	Characteristics
Auxin	IAA	Natural; light-sensitive
	NAA	Synthetic
	2,4-D	Synthetic. Has a high potency
Cytokinin	Zeatin	Natural
	Zeatin riboside	Natural
	Kinetin	Synthetic (prepared from DNA)
	BAP	Synthetic
Gibberellin	GA$_3$	Natural. There are over 50 different gibberellins based on the gibbane carbon skeleton. They are usually heat labile (i.e. cannot be autoclaved)

Table 4.2 Conjugates of plant hormones. The biochemical study of hormone-conjugating systems is still at an early stage and therefore definitive conclusions about the relative activities of free and conjugated growth regulators cannot always be given

Growth regulator class	Substance	Conjugated form	Activity
Auxin	IAA	IAA-glucoside	Inactive
		IAA-inositol	Active
		IAA-glucan	Uncertain
		IAA-inositol-glucoside	Uncertain
		IAA-glycoprotein	Uncertain
		IAA-aspartic acid	Active
	2,4-D	2,4-D-aspartic acid	Active
	NAA	NAA-aspartic acid	Active
Cytokinin	Zeatin	Zeatin riboside	Active
		Zeatin-glucoside	Active
Gibberellin	Many types	GA-glucoside	Inactive

This conceptual advance has at least allowed us to reconcile apparently incompatible experimental results, even if we are little nearer to defining the underlying molecular events. The important principles now recognized as influencing the relationship between added and perceived hormone concentrations are:

• Exogenous (and endogenous) hormones can be stored, modified or inactivated by the plant cell. For example, enzyme systems exist that conjugate auxins to sugars (giving hormonally inactive products) or to amino acids (hormonally active)

- There can be feedback control of synthesis of endogenous hormones by the added substances, as illustrated by the inhibition of abscisic acid (ABA) biosynthesis by high levels of ABA
- In cultures there can be carry-over (both intra- and extracellular) of hormones from one treatment to the next. An example of this behaviour is given by soybean cells that have been found to store added auxin as an amino acid conjugate that is not easily washed from the cells. On transference to an auxin-free medium, soybean can therefore still show behaviour (e.g. suppression of embryogenesis) that is characteristic of the presence of auxin
- Cultured cell clusters or tissues can regulate their internal hormonal environments. Cultured soybean cells have been shown to regulate their internal free auxin concentration to a constant value by conjugation of excess hormone.

For these reasons great care is needed in the interpretation of experimental results related to the concentrations of added plant hormones. However, a small number of hormone effects on differentiation do show a degree of consistency that make them useful guiding principles, even if they do not hold universally. These principles are:

- Relatively high auxin concentrations suppress organized growth and promote the formation of meristem-like cells
- The ratio of auxin to cytokinin influences the balance between root and shoot formation
- Cytokinins inhibit root formation
- Embryogenesis is stimulated by lowering the auxin concentration.

AUXIN:CYTOKININ RATIO

The idea that cultured tissues respond to the relative concentrations, rather than absolute amounts, of plant hormones was based originally on work with tobacco cultures, but has since been extended to many other species. It is not a universal phenomenon, however; for example, it does not apply in general to mono-cotyledons. Accepting these limitations, it is found that high auxin relative to cytokinin favours root formation and the converse situation favours shoot formation. This behaviour can be used to induce plantlet formation by sequential initiation of shoots and roots. It is not known how plant cells perceive auxin:cytokinin ratios.

THE NON-EQUIVALENCE OF HORMONES

Hormone classes (auxin, cytokinin, etc.) are defined on the basis of a small number of physiological tests (e.g. coleoptile curvature) usually employed for historical reasons. This classification has been developed arbitrarily, and does not correspond to sharp divisions of activity in the plant. Sometimes a given molecule may

Table 4.3   The non-equivalence of a range of auxins. It is easy to think of auxins (or other classes of hormone) as a single type of substance. However as these results from an experiment on aubergine tissue show, all auxins do not produce similar morphogenetic effects

Auxin treatment	Morphological effect
IAA	Shoot formation
NAA	Root formation
NOA	Callus formation
2,4-D	No effect on growth or organogenesis

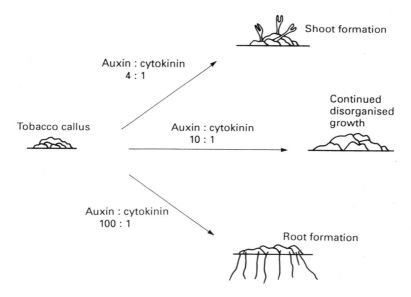

Fig. 4.4   The influence of auxin:cytokinin ratio on the morphological development of tobacco callus. This behaviour is shown in response to kinetin and IAA. The actual ratios vary for other types of auxin and cytokinin.

behave like one hormone (as defined in a physiological test) in some instances, and like another elsewhere. For example organogenesis is usually stimulated by auxins, but gibberellins can also fulfil this role in some cases. The non-equivalence of a variety of natural and synthetic auxins is shown in Table 4.3, and this further illustrates the difficulty that can be experienced when developing plant regeneration protocols.

Plantlet Formation from Cultured Tissue and Cells

A wide range of tissue-specific and environmental factors can affect the efficiency of a plant regeneration procedure, and these factors need to be optimized throughout the process. A large element of empiricism is also apparent here, and regeneration protocols essentially need adaptation for each particular species used.

There are two fundamentally different applications of plant regeneration technology. On the one hand it provides a rapid means by which plants can be propagated. In some cases, e.g. infertile hybrids, it represents the only method of replication. In these situations the aim is usually to produce plants that are genetically identical to their parents, i.e. it is a cloning process. Because of the genetic variation induced by disorganized growth during culture (Chapter 2), such growth must be avoided or at least minimized during clonal propagation. For this reason it is usual to regenerate plants from pre-existing meristems, e.g. axillary buds. The production of adventitious meristems can induce some degree of genetic variation. One reason for the genetic stability of existing meristems, or additional meristems produced by normal meristem development is that, because of the highly organized multicellular structure, single variant cells arising that lose compatability with the structure as a whole are unlikely to survive. In a disorganized structure, a variant cell can give rise to a complete variant meristem.

Plant regeneration technology also allows the recovery of plants from engineered or selected cells. In these cases the starting material may be a cell culture or individual protoplast. Genetic variation will therefore be evident and this may be disadvantageous if it leads to unavoidable downgrading of one set of plant characteristics during the optimization (by engineering or selection) of the intended set. Despite these unwanted effects of genetic variation in culture (somaclonal variation), the latter process can be exploited by using cell culture as a mutagenic treatment, followed by selection of enhanced genotypes that may arise (Chapter 5).

EXPLANT FACTORS

In cases in which the regeneration process is used for plant propagation, the normal starting material will be explant tissue from the species to be propagated.

The major requirements for effective explant tissue are a high cell division potential and morphogenic plasticity. These criteria are usually satisfied by immature, rapidly growing tissues from perennial species, but in the case of herbaceous plants, virtually any cells can be taken. There are exceptions to this rule, however. In the case of tulip, shoots can only be induced to form on lower stem explants from dormant bulbs. Usually, however, mature or highly organized tissue becomes morphogenetically 'determined'. This means that the cells are 'locked in' to a developmental programme, behaviour that limits the types of differentiation that can be experimentally induced. In certain conifers, this type of effect results in plagiotropic growth of regenerated shoots that produces excessive lateral branching and an unacceptable growth habit. The molecular basis of cell determination is not understood.

The growth conditions of the parent plant can have a large influence on the efficiency of regeneration from explant tissue. Lighting, humidity, and even seasonal effects are all important, and possibly exert their effect by modifying the hormonal status (e.g. the levels of conjugated, stored hormones) of the parent tissue.

Before explant tissue can be cultured, contaminating micro-organisms must be destroyed. Because micro-organisms divide so much faster than plant cells, they quickly overrun plant tissue culture media and usually kill the plant cells by direct competition for nutrients or the release of toxic compounds. Micro-organisms are usually confined to the surface of plant tissue and can be destroyed by surface sterilization, using sodium hypochlorite for example. Some micro-organisms however, invade the plant vascular tissue (e.g. in apple) and can be far more difficult to remove. In a few instances, plant cells and micro-organisms can coexist by metabolic co-operation, and infections can go undetected for long periods of time, only to show up when the composition of the growth medium is changed. The use of aseptically grown (e.g. shoot culture) material is a convenient way to circumvent sterility problems.

After selection of suitable explant material the next step is the culture of the tissue on the appropriate growth medium. Solidified media are usually used to prevent waterlogging of the tissue (pp. 94–5). The composition of plant tissue culture media has been discussed in Chapter 1. A problem frequently encountered in the initial stages of culture is the browning and eventual death of the tissue due to the excessive production of polyphenols, possibly by the triggering of defence reactions. The problem can sometimes be alleviated by the incorporation of adsorbents, for example activated charcoal or polyvinylpyrrolidone, or an antioxidant such as ascorbic acid. The inclusion of adsorbents must be carefully controlled, however, because they can cause depletion of media components.

The type of regeneration that is being attempted will determine the optimum hormone regime to be imposed. Propagation by axillary meristem multiplication requires a relatively high cytokinin level to suppress dominance effects. The induction of adventitious meristems is usually dependent on a high auxin treatment. In some cases, e.g. with nodal tissue, the material is sufficiently organized to produce and distribute hormones endogenously, and plant regeneration occurs in the absence of added growth regulators.

Explants containing vascular tissue (stem segments for example) may exhibit different developmental effects depending on their orientation relative to the nutrient agar. This type of effect is thought to be due to the intrinsic polarity, with respect of hormone transport, existing in the tissue. We shall consider this process in relation to auxins, but a similar situation may exist for other hormones. Many auxin-mediated physiological responses (e.g. phototropism) are co-ordinated by the polar (unidirectional) transport of the hormone through the tissue. In shoots, indoleacetic acid is transported from the tip to the base. Transport is transmembranal and probably occurs through the cells of the vascular cambium. Auxin movement is controlled by transport proteins located at the basal ends of the cambium cells. Explants containing vascular tissue therefore possess an intrinsic polarity, and this has been clearly demonstrated by an investigation of the effect of an inhibitor of polar auxin transport, tri-iodobenzoic acid (TIBA), on bud formation in tobacco stem explants. On the appropriate medium, in the absence of TIBA, buds were formed only at the basal end of horizontally placed explants. In the presence of TIBA, however, the number of buds was reduced, but those developing did so over the entire surface of the explant. Clearly, tissue placed on a solidified medium is exposed to artificial hormone (and nutrient) gradients because only one side of the tissue is in contact with the growth medium. These gradients can either reinforce or oppose the intrinsic tissue polarity. There are numerous reports of the effect of explant orientation on subsequent development of the tissue. For example, dandelion root segments placed in their natural (i.e. as in the plant) orientation on nutrient agar produce shoots from their upper cut surface. If the tissue is inverted, disorganized callus forms at the upper surface instead.

THE INDUCTION OF ORGANIZED GROWTH IN CULTURED CELLS

The major prerequisite for plantlet regeneration from cultured cells (including protoplast-derived cells) is the induction of rapid cell division. This is usually achieved by exposing the cells to a high auxin concentration, in the presence of cytokinin. This type of treatment frequently results in tightly aggregated clusters of small, meristematic-like cells (meristemoids) in which further development can be induced by the appropriate hormonal stimuli. The process is analogous to adventitious meristem formation in explant tissue.

PLANTLET DEVELOPMENT

Following meristem induction or multiplication, plantlet development is similar whether the starting material was of whole plant, or cell culture origin. Shoot, followed by root, development is induced, usually by manipulation of the auxin: cytokinin ratios as discussed previously. Environmental factors, e.g. lighting, temperature, need to be optimized for efficient plantlet production. A relatively sharp optimum of around 1000 lux has been reported for shoot formation from a range of species. In addition the photoperiod can be an important variable, as in the case of *Vitis* in which a short-day treatment is required for shoot induction. It is usual to increase the light intensity gradually as the regenerated plantlets mature.

A temperature of 25°C is usually used in regeneration experiments and appears to be near optimum for most species, although certain tropical plants, e.g. date palm, require higher temperatures, in the range 27–30°C.

VITRIFICATION

A fairly common problem encountered during plantlet regeneration is the irreversible swelling and distortion of the regenerating tissues, accompanied by translucence, and sometimes leading to tissue death. This process is termed *vitrification*. It seems to result from waterlogging of the tissue and consequently is seen more frequently in liquid than on solid media. For this reason, media solidified with agar are almost universally used for plant regeneration. Vitrific-ation may be a consequence of the abnormally low levels of wax found in the cuticles of regenerated plant tissues. Low wax deposition may result in some way from the high humidity usually existing inside closed culture vessels, or it may be due to inhibition of wax biosynthesis by the hormonal regimes necessary for plant regeneration. Another consequence of tightly closed culture vessels is the build-up of ethylene which can suppress or arrest growth. Vessel tops are therefore designed to be loose fitting and so allow a certain degree of gas exchange without increasing the risk of contamination by the influx of micro-organisms.

ACCLIMATIZATION OF REGENERATED MATERIAL

The lack of fully functional cuticle in regenerated plant tissues means that plants transferred directly from a culture vessel to a normal outside environment will often die due to excessive water loss. It is necessary therefore to acclimatize newly regenerated plants slowly to the normal growth conditions, during which time there is a build-up of cuticular wax. Acclimatization is achieved by covering the potted-out plants with polyethylene bags and either removing the latter for progressively longer periods, or by punching an increasing number of holes in them.

DECLINE IN MORPHOGENETIC POTENTIAL

It is almost universally observed that cell or tissue cultures progressively lose their ability to undergo total regeneration, or even limited differentiation, as they are

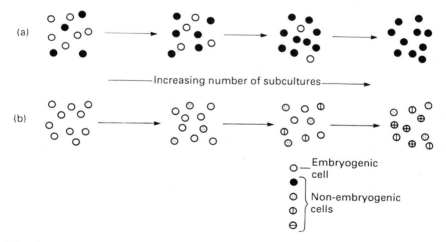

Fig. 4.5 Models for the loss of embryogenic potential of cultured plant cells. (a) Embryogenesis due to competent cells present in the original explant. These cells are at a competitive disadvantage under culture conditions and are eventually diluted out. (b) Mutations result in the loss of the ability of the affected cells to undergo embryogenesis. With increasing duration of culture, the majority (or all) of the cells become so mutated. Note that these models are not mutually exclusive.

maintained for longer periods in culture. In some cases this decline can be extremely rapid (e.g. over one subculture) or relatively slow (e.g. over several years). Two hypotheses have been put forward to explain this decline. On one model, morphogenesis results from differentiation of competent cells that are direct descendants of the population of competent cells that were present in the explant. If these cells are at a competitive disadvantage, or if a variant, non-morphogenic cell type arises that is better adapted to growth in culture, the morphogenetically competent cells will eventually be diluted out of the population. The second hypothesis states that morphogenetic ability is lost as a result of progressive accumulation of mutations that prevent differentiation. These two models are not mutually exclusive, and both of these mechanisms probably operate to some extent. An example of a genetic or epigenetic change that occurs in cultured cells and can inhibit morphogenesis is the enhancement of endogenous auxin production. High levels of auxin favour disorganized growth. Furthermore, auxin synthesized in the overproducing cells can diffuse throughout the culture and suppress morphogenesis even in 'wild type' cells.

REGENERATION OF *AGROBACTERIUM*-MEDIATED TRANSFORMANTS

The intense research into the use of the plasmids of *Agrobacterium tumefaciens* and *A. rhizogenes* as gene vectors has focused attention on the special characteristics of *Agrobacterium*-transformed cells with respect to plant regeneration. Such cells exhibit behaviour that causes problems in regeneration, but that also offers unique opportunities for the study of the mechanisms of growth control and differentiation-related processes in plants. The transformation event with *Agrobacterium* is the insertion of certain genes from a transferred plasmid into the host cell genome (see Chapter 5). The inserted DNA includes genes that modulate the auxin and cytokinin levels in the host or the efficiency of the cellular perception of these hormones. Transformation by wild type *A. tumefaciens* induces galls that consist of outgrowths of undifferentiated, callus-like tissue. This transformed tissue cannot be induced to regenerate into plantlets by manipulation of the external hormonal regime, presumably due to inhibitory levels of hormones produced endogenously. Clearly this behaviour severely limits the use of wild-type *A. tumefaciens* as a gene vector for the production of engineered plants. The regeneration problem has been overcome by deleting the plasmid genes that code for the hormone-associated functions. During experiments to delete these functions, interesting developmental effects were noted. When the auxin-associated loci were deleted, the resulting gall tissue showed a tendency to form shoots; conversely when the cytokinin locus was deleted, roots were formed. These experiments strongly support the auxin:cytokinin ratio model for the control of organogenesis (pp. 90–91). More precise manipulation of the hormone-associated functions of *A. tumefaciens* may prove to be an extremely useful experimental approach to the elucidation of hormone action. Transformation by *A. rhizogenes* results in an outgrowth of roots (hairy root disease) from transformed tissue, which can also be induced to undergo regeneration into plants. The *A. rhizogenes* root-inducing effect is also due to the balance of hormones (high auxin:cytokinin ratio) induced in the transformed cells. Hairy roots can be maintained indefinitely in culture, and may be a useful source of secondary metabolites (Chapter 6).

The Biotechnological Application of Regeneration Techniques

We have seen that regeneration technology can be used to recover improved plant varieties from engineered and selected cells (see also Chapter 5), and is also employed in the clonal propagation of plants of agricultural and horticultural importance (micropropagation). Plant regeneration techniques also find a use in other areas of biotechnology such as the production of pathogen-free plants and germplasm storage. We now briefly consider the commercial operation and status of these procedures.

REGENERATION FROM ENGINEERED AND SELECTED CELLS

Regeneration is a necessary step in the recovery of plants from cells arising from engineering, selection (Chapter 5), or cell fusion (Chapter 3) procedures. Often it is plant regeneration that is the limiting step in the overall process and this reflects our lack of understanding of, and therefore control over, plant developmental

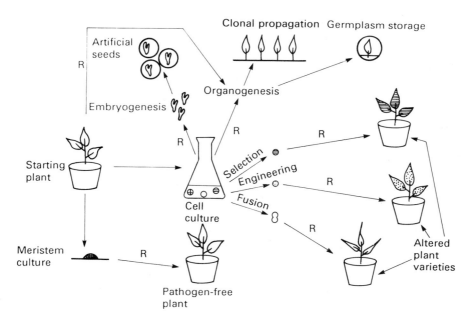

Fig. 4.6   The biotechnological exploitation of regeneration techniques. The steps marked 'R' rely on the regeneration of viable plants from cells or explants.

pathways. Up to the present, extremely few of the new plant varieties that have been produced by these techniques are commercially relevant. Some of these are described in Chapters 3 and 5. Traditional plant breeding remains by far the most powerful means we have for the production of improved plants, although this situation may change rapidly as we become able to apply the new technologies, as a result of a better understanding of plant systems.

CLONAL PROPAGATION

The great commercial success story of regeneration technology is plant propagation, and many ornamental and agricultural plant species are routinely multiplied by these techniques. Micropropagation is most important for the multiplication of infertile plant hybrids, for example orchids, or for highly heterozygous (variable) species such as potato and strawberry, for which clonal propagation provides a means of producing uniform plants. Another advantage of micropropagation is the absence of seasonal constraints, i.e. propagation can be continued all year round.

Micropropagation is efficient but labour intensive, and therefore expensive. The process of regeneration requires many sterile transfers and media changes, and whilst labour-saving mechanization of these procedures is being explored, human operators are proving difficult to supersede because much judgement is required concerning the best tissue to transfer, and the optimum timing of the various steps.

Somatic embryogenesis is in principle more efficient in terms of speed of plantlet production than separate root and shoot initiation because large numbers of embryos can be formed from a single explant. The process requires fewer media transfers and results in material that is of more uniform size and shape; consequently it presents a better target for mechanization. Somatic embryogenesis has been demonstrated in a variety of commercial plant species including coffee, cotton, celery, *Citrus*, grape, date palm, and millet. The main disadvantage of embryogenesis is that, because it involves an initial step of disorganized growth during which the pre-embryonic, meristem-like cells are formed, a high degree of genetic instability is usually induced. The propagated plants are therefore not necessarily genetically identical to the parent. For this reason, somatic embryogenesis has not been used commercially for 'clonal' propagation except in a very few cases, e.g. oil palm. In this instance the species is normally propagated by seed, which, because of the highly heterozygous nature of the plant, results in very variable offspring. The oil palm plant has a single meristem, and there is no known 'natural' method for vegetative propagation (e.g. taking cuttings), which might produce clonal offspring. Given the variability of seed-produced plants, a degree of culture-induced variation (somaclonal variation) is acceptable.

An additional, rather speculative use for embryogenesis is in the production of artificial seeds. The idea here is that species propagated by somatic embryogenesis might be planted out directly at the embryo stage so that the costly process of *in vitro* development could be avoided. Small-scale experiments on species including alfalfa, cauliflower, and celery, have been performed in which somatic embryos have been encapsulated in an artificial coat (often calcium alginate) containing the nutrients and hormones supplied by the seed endosperm in nature. Before this technique can become a commercial reality, problems of synchronization of the initial stages of somatic embryogenesis, and of mechanization of repetitive parts of the process need to be overcome because of the extremely time-consuming nature of the technique in its prototype form. The method would only prove acceptable if culture-induced variation could be minimized or abolished, or in special cases in which other propagation methods resulted in higher variability of derived plantlets.

In principle the propagation of tree species is an especially favourable target for *in vitro* regeneration techniques. Trees are frequently highly heterozygous, with long life cycles, and controlled sexual crossing is extremely difficult. Unfortunately, micropropagation is only usually possible from juvenile specimens, which are too young to be assessed for suitability as breeding stock. The reasons for the failure of tissues from mature trees to undergo regeneration are largely unknown, but tissue determination (pp. 92–3) is probably involved. In some cases mature trees can be induced (e.g. by repeated pruning) to send out new growth which can be amenable to regeneration techniques. Because of the difficulties associated with the conventional breeding and clonal propagation of elite trees, advances in tree micropropagation are likely to have a profound effect on forestry practice.

THE PRODUCTION OF PATHOGEN-FREE PLANTS

In crop species that are routinely propagated vegetatively (e.g. potato) there is usually a severe risk of passing on systemic viral infections (e.g. potato virus x, potato leafroll virus) during the propagation process. Viruses can thus rapidly spread and affect a very large proportion of a particular crop, often with devastating results. Other pathogens can be transmitted during micropropagation but because of the intimate, intracellular association of viruses with plant tissue, viruses constitute by far the largest threat to vegetatively propagated crops. There is therefore much effort being expended in the virus testing (usually by serological methods) of plant material prior to propagation, and also the development of *in vitro* methods of curing infected plant material of viruses. Generally speaking the design and application of antiviral chemicals to plants has proved rather ineffective because the nature of plant–viral interactions is poorly understood. The most important means of virus eradication at present is that of meristem isolation and regeneration into plants (meristem tip culture). For reasons that are still unclear the majority of plant viruses are unable to enter or survive in meristem cells. It is probable that the lack of vascular connections to plant meristems exclude phloem- and xylem-borne viruses. There is also some evidence that certain viruses that do enter meristem cells are destroyed.

Meristem tip culture consists of excision, and growth on the appropriate medium of the apical 0.5–1 mm of shoot tips. Heat treatment and chemotherapy can also be applied to increase the chance of total virus eradication. Usually only a small percentage of the meristem tips survive. This percentage can be increased if larger sections of tissue are taken, but there is a corresponding increase in the risk of incomplete virus destruction. Once freed from virus, sterile shoot cultures can provide an indefinite source of clonal, virus-free material.

GERMPLASM STORAGE

In a global context, germplasm storage is a vitally important enterprise. Modern agricultural practices tend to reduce the genetic diversity of the plants utilized. It is of the utmost importance that we conserve agriculturally 'outmoded' species, for example the wild relatives of cultivated species, as a reservoir of genetic characteristics, e.g. disease resistance, that will be available in the event of crop losses due to environmental changes, increasing pathogen virulence, etc. Seed banks and cryopreservation schemes are being set up with these eventualities in mind.

Germplasm storage is also important at the level of the individual plant and it is here that regeneration techniques come into play. Certain individual plants exhibit a combination of characteristics that make them especially suitable for commercial use. For example so-called 'elite' trees may show straight growth and a relative absence of branching. Because of the variable nature of tree seed, elite individuals cannot be reliably propagated sexually. However, by the use of micropropagation techniques in conjunction with long-term *in vitro* culture, individual genotypes can be indefinitely maintained. Long-term culture is usually achieved by holding shoot cultures at 4 °C to prolong the growth period, and minimize the need for labour-intensive subculturing.

References and Further Reading

Burgess, J. (1985). *An Introduction to Plant Cell Development*. Cambridge University Press, Cambridge. (Clearly presented information on the cell biology of development in plant cell tissues.)

Evans, D.A., Sharp, W.R., Ammirato, P.V., and Yamada,Y. (eds) (1983). *Handbook of Plant Cell Culture*, Vols 1–4. Macmillan, New York. (This extensive reference work gives details of general methods and application to particular species.)

George, E.F. and Sherington, P.D. (1984). *Plant Propagation by Tissue Culture*. Exegetics, Basingstoke. (The classic text on micropropagation.)

Haissig, B.E., Nelson, N.D., and Kidd, G.H. (1987). Trends in the use of tissue culture in forest improvement. *Biotechnology* **5**, 52–59.

Levin, R., Gaba, V., Tal, B. *et al.* (1988). Automated plant tissue culture for mass propagation. *Biotechnology* **6**, 1035–1040.

Martin, C. (1985). Plant breeding *in vitro*. *Endeavour* **9**, 81–86.

Murashige, T. (1974). Plant propagation through tissue cultures. *Annual Review of Plant Physiology* **25**, 135–166.

Redenbough, K., Paasch, B.D., Nichol, J.W. *et al.* (1986). Somatic seeds: encapsulation of asexual plant embryos. *Biotechnology* **4**, 797–801.

Yeoman, M.M. (ed.) (1986). *Plant Cell Culture Technology*. Blackwell Scientific Publications, Oxford. (Contains individual chapters on vegetative propagation and tissue culture by G. Hussey, W.R. Scowcroft, and S.A. Ryan.)

Chapter 5

Improvement of Plants via Plant Cell Culture

RAY CRESSWELL

Introduction

The techniques of plant cell culture facilitate the rapid production of variant cell lines via selection procedures similar to those employed in microbial systems. These variant cell lines are useful in research into the genetics and biochemistry of plant cells and in biotechnology for production of new plant varieties and secondary metabolites. We consider here some of the principles and problems associated with the selection of variant plant cells and the regeneration of plants with new characteristics. Later we see how the application of genetic engineering methods is now aiding this area of research.

SELECTION OF VARIANT CELL LINES

There are a number of requirements for the selection of variant cell lines from populations of cultured plant cells. These include:

- Suitable cells or tissues with established and reliable culture conditions for their growth
- A source of genetic, or epigenetic, variation among the cells which may either be spontaneous or induced artificially
- A protocol for the selection and maintenance of the variant cells.

Each of these requirements merits some consideration here.

Plant cell selection has been performed with equal success using cells growing on solid media as calluses and in suspension as typical cell clumps (see Chapter 1).

Table 5.1 Some examples of selected variant cell lines with tolerances to antimetabolites

Species	Mutagenic treatment	Antimetabolite
Datura innoxia	NSG[a] and others	5-Methyltryptophan
Datura innoxia	None	Aminopterin
Nicotiana tabacum	None	p-Fluorophenylalanine
Daucus carota	None	Azetidine-2-carboxylic acid
Nicotiana sp	None	Streptomycin
Nicotiana tabacum	ENU[c]	Chlorsulfuron
		Sulfometuron methyl
Nicotiana tabacum	EMS[b]	Methionine sulfoximine
Nicotiana tabacum	None	5-Bromodeoxyuridine
Daucus carota	None	5-Methyltryptophan
		p-Fluorophenylalanine
		Ethionine
		Aminoethylcysteine
Petunia hybrida	None	Methotrexate
Acer pseudoplatanus	None	p-Fluorophenylalanine
Medicago sativa	EMS[b]	Ethionine
Nicotiana tabacum	EMS[b]	Cycloheximide

[a] *N*-methyl-*N'*-nitro-*N*-nitrosoguanidine
[b] Ethyl methane sulphone
[c] Ethyl nitrosourea

Rapidly growing, fine suspension cultures or friable calluses are generally the most suitable for selection purposes. Occasionally the cell clumps can be broken down to form single cells, usually by protoplasting (see Chapter 3), in order to allow selection protocols analogous to those used with microbial cultures. Often however, such procedures are unnecessary.

The genetic instability of cultured plant cells (see Chapter 2) frequently produces sufficient variants within a population without the need for artificial mutagenic treatments. Mutagenic agents such as ultraviolet light, ethylmethane sulphonate, and nitrosoguanidine are sometimes employed to increase the frequency of variants (Table 5.1) but evidence for their efficacy is uncertain.

A variety of protocols has been used in order to isolate variant plant cells and the method of choice depends on the nature of the cell culture and the type of variant to be isolated. The selection procedures described here serve as examples only and do not form an exhaustive account of all the approaches employed. There have been many reports of selection of plant cell lines exhibiting auxotrophy or resistance to antimetabolites (Table 5.1). Auxotrophic variants have not been so readily applied to research on plant improvement and are not considered here. Antimetabolite-resistant variants, however, are among the simplest to isolate and, as we shall see later, provide useful model systems towards the production of improved plants through *in vitro* selection. The aim of a selection programme for the isolation of antimetabolite-resistant cells is to provide sufficient toxin, incorporated in the medium, to kill all normal cells whilst allowing resistant variants to grow. For callus cultures on solid media the tissues are incubated, usually in the presence of toxic or semitoxic concentrations of the antimetabolite, for periods of several months. The effects of these treatments are either to allow only resistant cells to grow, or to produce a gradual loss of the sensitive cells and a domination of the cultures by resistant variants. It is conceivable that during selection a single resistant cell may grow to form a new resistant culture. However, such single-cell growth does not appear to require the usual special conditions, e.g. inclusion of conditioned medium (see Chapter 1), most likely because the dying sensitive cells supplement or 'condition' the medium themselves.

Similarly, plant cells can be selected for antimetabolite-resistance in suspension culture. Procedures for this vary but two approaches provide useful examples. Suspension cultures may be treated with toxin at the minimum concentration that causes almost total cell death. The apparently 'dead' cultures are then incubated for 2–3 months in order to allow surviving resistant cells to grow. Such resistant cells often first become visible as small, viable white/cream cell clumps among the otherwise dead cells and eventually grow to dominate the culture. Alternatively, cells in suspension may be subjected to low doses of toxins, which kill a proportion of the population, and doses are increased whenever the cells achieve sufficient growth to allow their subculture. With such an approach it is sometime possible to produce cells which grow with markedly toxic levels of antimetabolites after just a few (e.g. 4–5) subcultures. Whichever method is chosen, once the cultures become fully established they should be capable of growth in the presence of antimetabolite concentrations highly toxic to unselected cells.

It may then be necessary to subject the cultures to continuous selection pressure in order to maintain the selected characteristics. For example, methotrexate-resistant *Petunia hybrida* cells were found to lose their resistance during growth in the absence of methotrexate. Similarly, cycloheximide-resistant *Nicotiana tabacum* cell lines were shown to have transient resistance, lost when the drug was omitted from the medium. Such transient resistance can often be attributed to alterations in gene expression (or other epigenetic events) rather than mutations (see Chapter 2). Other resistant cell lines, however, have been found to be stable in the absence of selection pressure. Thus, for example, *N. tabacum* cells resistant to the amino acid analogues, 5-methyltryptophan and *p*-fluorophenylalanine retained their resistances in the absence of the antimetabolites.

Frequently, it is difficult to demonstrate unequivocally that selected altered phenotypes are due to true mutational events. In such a situation it is preferable to refer to selected resistant cells as variants rather than mutants. Demonstration of a true mutant may require crossing of plants regenerated from selected cell lines and analyses of progeny (p. 34).

PRODUCTION OF VARIANT PLANTS FROM SELECTED CELLS

Where it is possible to regenerate plants from variant cells (see Chapter 4) selection techniques have potential for production of crop varieties with new characteristics

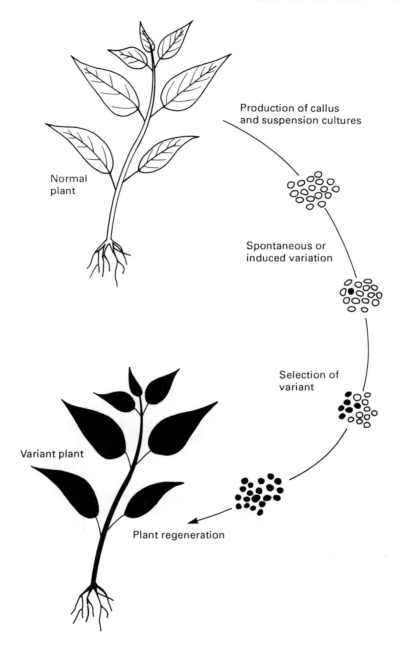

Fig. 5.1 Production of variant plants through selection in cell culture.

some of which, e.g. herbicide and salt tolerance, may be desirable in field conditions (Fig. 5.1). This section describes some examples but for further information see the reviews by Maliga (1984), Chaleff (1981) and King (1984).

Some of the most thoroughly studied variant cell lines are those with resistances to amino acid analogues and antibiotics. The mechanisms of resistance in many of these cell lines have been elucidated, and although such cells may seldom have practical use in production of plants for agriculture, they have proved very instructive for related research. We shall consider some of these cell lines here.

The tryptophan analogue 5-methyltryptophan usually causes death of cultured plant cells, but a number of resistant cell lines have been selected and studied. Tryptophan synthesis occurs at the end of the shikimic acid pathway (Fig. 5.2) and is controlled by feedback inhibition of anthranilate synthetase by tryptophan. Analogues of tryptophan have been shown to act by inhibiting anthranilate synthetase in cell cultures. 5-Methyltryptophan-resistant cells, however, frequently contain a form of anthranilate synthetase less sensitive to inhibition by the analogue and tryptophan itself. Such variants can be shown to accumulate

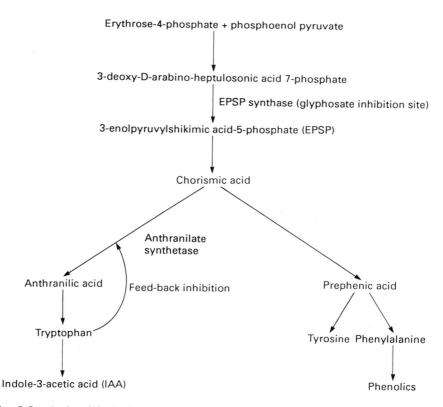

Fig. 5.2 A simplified diagram of the shikimic acid pathway.

elevated levels of free tryptophan and grow in the presence of the analogue. When whole plants were regenerated from the resistant cells they also were found to be resistant to 5-methyltryptophan and contained increased tryptophan levels, and anthranilate synthetase activity less sensitive to the analogue. Furthermore, callus reinitiated from the regenerated plants displayed similar characteristics. The resistance phenotype selected in cultured cells was, therefore, maintained and expressed at the whole plant level. Variant plants regenerated in this way from selected cells must be analysed genetically in crossing experiments to determine the heritability of the new trait (Fig. 5.3). This approach to the production of new variant plants, however, is sometimes hindered by failure of selected cells to regenerate plants or failure of regenerated plants to flower normally. Some 5-methyltryptophan-resistant *Daucus carota* cells, for example, were found to regenerate poorly. These cells contained anthranilate synthetase less sensitive to feedback inhibition by tryptophan and over-accumulated tryptophan and indole-3-acetic acid (IAA) (see Fig. 5.2). It was postulated that the elevated IAA levels in the cells reduced their capacity to regenerate. There may, however, be other reasons why selected cells fail to regenerate, e.g. loss of totipotency during prolonged culture (see Chapter 4).

For production of variants it is sometimes preferable to select cells from a haploid population derived from cultured pollen, especially if the selected trait is likely to be recessive. The haploid cells are then made diploid later by treatment with colchicine or spontaneously by endoreduplication. This approach, however, may result in some loss of hybrid vigour in the regenerated variant plants.

Several plant cell lines have been selected with resistances to antibiotics such as cycloheximide, kanamycin and streptomycin. Cycloheximide-resistance in a tobacco cell line was shown to be associated with differentiation of the cells because resistant callus was always of a shoot-differentiating type. Loss of resistance was accompanied by loss of shoot formation. The resistance, which was attributed to conversion of cycloheximide to a biologically inactive form, was thought to be due to activation of a gene (or genes) not expressed in non-differentiating callus. Such differences in gene expression between differentiated and non-differentiated plant tissue can complicate attempts to regenerate variant plants from variant cells. Characteristics selected in tissue culture, for example, may not always be expressed at the whole plant level following regeneration. Cultured cells may undergo epigenetic changes, not due to true mutations, which are lost on regeneration.

Some selected characteristics may be determined by the chloroplasts and/or mitochondria of the cell rather than the cell nucleus. An illustration of this, which was also one of the first examples of the regeneration of variant plants from selected cells, was provided by a streptomycin-resistant *N. tabacum* line. Haploid callus was selected for streptomycin resistance and regenerated into plants which were found to be diploid (through diploidization of haploid cells) and streptomycin-resistant. Crossing experiments with the regenerated plants indicated a uniparental, non-Mendelian inheritance attributed to resistance being carried by the chloroplasts or mitochondria.

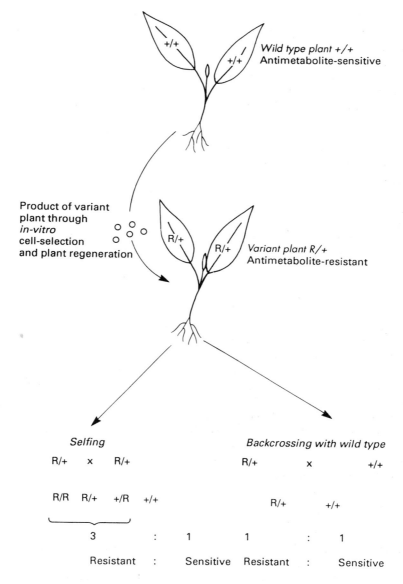

Fig. 5.3 Genetic analysis of regenerated variant plants. Regenerated variant plants with resistance to antimetabolites are analysed genetically through crossing experiments and progeny examination. In this example a dominant allele R, selected *in vitro* and conferring tolerance to an antimetabolite, yields resistant and sensitive progeny at ratios of 3:1 and 1:1 respectively after selfing and backcrossing.

Table 5.2 Herbicide-tolerant cell lines and regenerated plants

Herbicide	Species of selected cell line	Tolerance retained in regenerated plants?
Picloram	*Nicotiana tabacum*	Yes
2,4-D	*Trifolium repens*	N/A
2,4,5-T	*Trifolium repens*	N/A
Glufosinate	*Medicago sativa*	N/A
Sulphonylureas	*Nicotiana tabacum*	Yes
Imidazolinone	*Zea mays*	Yes
Glyphosate	*Corydalis semperivens*	N/A
	Petunia hybrida	N/A
	Nicotiana tabacum	Yes

N/A: data not available

PRODUCTION OF HERBICIDE-TOLERANT PLANTS

Plant cell cultures have assumed an important role in the study of the mode of action of some herbicides. In addition there are now numerous examples (Table 5.2) where the techniques of *in vitro* selection and plant regeneration have been used to produce herbicide-resistant cells and plants. Production of new crop varieties in this way may have important applications in agriculture by allowing the effective control of weed species in a field without damage to the crop.

Chaleff and Parsons (1978) selected cultured tobacco cells with tolerance of the herbicide picloram and then regenerated plants from four of the tolerant cell lines. The plants were subjected to crossing experiments in order to examine inheritance of the new trait among the progeny. When the plants were backcrossed and self-fertilized three of the cell lines produced resistant and sensitive progeny at ratios of 1:1 and 3:1 respectively. Self-fertilization of the fourth cell line, however, produced semi-tolerant, tolerant, and sensitive progeny at ratios of 2:1:1. Three of the mutations, therefore, behaved as dominant alleles and one behaved as a semi-dominant allele, of single nuclear genes. Analyses of this type are necessary in order to provide information on the genetics of newly selected traits particularly if the new characteristic is to be incorporated into crop varieties.

Imidazolinones are a class of herbicides which inhibit acetohydroxyacid synthase (AHAS), the first enzyme on the biosynthetic pathway of the amino acids valine, leucine, and isoleucine. Exogenous supplies of the three amino acids have been shown to prevent the effects of the herbicides on maize tissue cultures and seedlings. Shaner and Anderson (1985) selected imidazolinone-tolerant cultures of maize by continuous incubation of callus cultures in the presence of the herbicide. Plants were then regenerated from the selected callus tissue and raised to produce pollen which was crossed with inbred maize. Of the progeny produced from the crossings, 41% showed increased tolerance of the herbicide. When AHAS activity from the tolerant cell lines and plants was assayed it was found to be much less sensitive to imidazolinone class herbicides than AHAS activity from unselected tissue. This work has now yielded maize plants which remain undamaged when sprayed with imidazolinone treatments lethal to unselected maize.

The herbicide glyphosate (N-[phosphonomethyl] glycine) inhibits the shikimic acid pathway enzyme 5-enolpyruvylshikimic acid-3-phosphate (EPSP) synthase (see Fig. 5.2) and so prevents synthesis of aromatic amino acids. Treatment of cultured cells with the herbicide has been shown to result in accumulation of the EPSP precursor, shikimic acid-3-phosphate (and/or shikimic acid), a reduction in levels of free aromatic amino acids and eventually, death. A number of glyphosate-tolerant cell lines have been selected and characterized. EPSP synthase activities in the tolerant cells have been shown to be as much as 30–40 fold higher than those in unselected cells. However, in contrast to imidazolinone-tolerant cultures which produced a herbicide target enzyme with reduced sensitivity to imidazolinones, the sensitivities of EPSP synthases to glyphosate in glyphosate-tolerant and unselected cultures were similar. Recent evidence suggests that glyphosate tolerance in selected plant cells is brought about by marked overproduction of the normal EPSP synthase enzyme due to an increase in the number of copies of the relevant gene (i.e. gene amplification, see Chapter 2). A similar mechanism has been shown to be present in plant cells with resistance to the non-selective herbicide glufosinate (often referred to as phosphinothricin). Selected alfalfa cells which were 20–100 times more resistant to glufosinate than unselected cells were shown to have 3–17-fold increases in levels of the target enzyme, glutamine synthetase (GS) and a 4–11-fold amplification of the GS gene. Such a mechanism of tolerance, whilst being very effective in cell culture, may be difficult to maintain in plants regenerated from the cells and indeed with glyphosate tolerant lines; for instance, there are no clear examples to date where this has been done (see Chapter 2 for a discussion of the generation and loss of amplified genes). Later, therefore, we will examine how this problem is now being circumvented with the aid of molecular biology.

In a novel approach, Radin and Carlson (1978) isolated herbicide-tolerant tobacco cells by direct selection of tissue from leaves. γ-Irradiated haploid plants were sprayed with bentazone and phenmedipharm, herbicides which bleach leaves but do not affect callus cultures of tobacco. Green, herbicide-tolerant sectors which survived the treatments were then excised from the surrounding dying leaves, cultured, and regenerated as diploidized plants. The regenerated plants were crossed with normal plants but the F_1 progeny were found to be herbicide-sensitive. When, however, the F_1 progeny were self-fertilized they produced herbicide-tolerant F_2 progeny at ratios to sensitive progeny which indicated the presence of recessive mutations (Fig. 5.4).

PRODUCTION OF CHILL-TOLERANT PLANTS FROM CULTURED CELLS

There are a number of accounts of selection of cultured plant cells with enhanced tolerance of chilling, but descriptions of plants regenerated from such cells are rare.

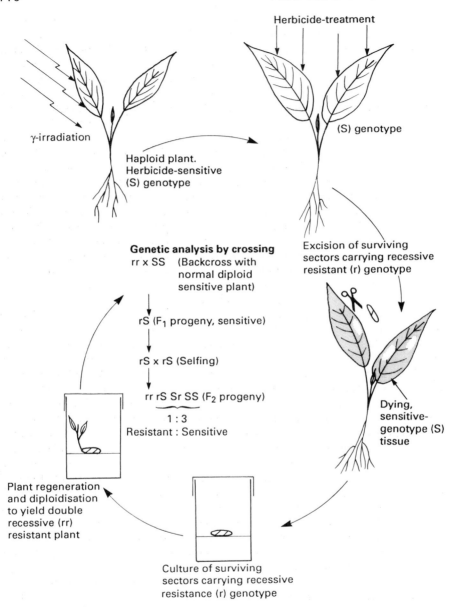

Fig. 5.4 *In situ* mutant selection. Haploid, herbicide-sensitive plants were γ-irradiated and treated with herbicide. Surviving (resistant) tissue sectors were then excised, cultured, and regenerated into diploidized plants. Genetic analysis by crossing experiments demonstrated the presence of recessive mutations conferring herbicide tolerance in the selected tissues (Radin and Carlson, 1978).

It seems likely that the lack of success in this area reflects the greater complexity of the physiological changes (e.g. alterations in membrane lipid composition) required in order to confer chilling-tolerance upon the cells. Despite the difficulties, Dix and Street (1976) were able to select cell lines of *Nicotiana sylvestris* and *Capsicum annum* with enhanced ability to recover from chilling after exposing the cells to low temperatures ($-3°-+5°C$) for 21 days. Unfortunately, callus cultures derived from the sexual progeny of plants regenerated from the chilling-tolerant *N. sylvestris* cells proved to be as sensitive to chilling as unselected cultures. It was suggested that either the tolerance was epigenetic and lost on regeneration or that the selected callus contained a mixture of tolerant and sensitive cells from which the sensitive cells were more amenable to regeneration. One notable success in this area was the production of cultivars of *Euphorbia pulcherrima* with reduced leaf fall under cold conditions after X-irradiation of suspension cultures and exposure to low temperatures (12 °C) for 170 days. These and related studies on the responses of plant cells to low temperatures may provide useful information relevant to research on the long-term cold-storage (cryopreservation) of plant cells. Nevertheless, it is clear that for *in vitro* selection to be used effectively in the production of chilling-tolerant plants more research is required.

PRODUCTION OF SALT-TOLERANT PLANTS THROUGH *IN VITRO* SELECTION

The productivities of many agricultural areas are limited by the accumulation of salt in the soils (especially in warm climates where water loss through desiccation is rapid) and this has led to considerable interest in the development of salt-tolerant plants through *in vitro* selection. The relatively high productivities of many naturally occuring salt-tolerant species has encouraged some enthusiasm for this approach. When considering selection of this kind it is important to remember that salt-tolerance may be conferred upon plants in a variety of ways some of which may be difficult to select for *in vitro*. Thus, for example, callus cultures isolated from some highly salt-tolerant plants have been found to be more sensitive to salt than callus produced from sensitive varieties. In such instances it appears that tolerance to salt is due to adaptations at the level of the whole plant rather than the cell. Others, however, have found somewhat greater similarities in response to salt between callus cultures and the plants they were derived from. Examples of plausible tolerance mechanisms include salt exclusion or compartmentalization and/or production of organic solutes such as proline which help maintain osmotic balances in saline environments. Some salt-tolerant cell lines are also known to produce new or enhanced quantities of polypeptides which are thought to have a role in the tolerance mechanism.

A number of salt-tolerant cell lines have been selected from cultures incubated in the presence of, for example, 0.5–1% NaCl. Interestingly, these selected cells sometimes display a halophytic response and achieve only poor growth in the absence of salt. For some species, e.g. *Nicotiana sylvestris*, *Datura innoxia*, and *Daucus carota*, there is evidence that the selected salt tolerance is retained in plants regenerated from the cells. Limited data suggest that the tolerance may sometimes, but not always, be transmitted to subsequent generations derived from the regenerated plants. However, plants regenerated from salt-tolerant cell lines of some other species have been found to be sensitive to salt treatment.

PRODUCTION OF DISEASE-RESISTANT PLANTS THROUGH *IN VITRO* CULTURE

The extent to which *in vitro* selection can be used for production of disease-resistant plants is determined somewhat by the nature of the host–pathogen relationships available for research. Many are rather complex interactions, physically and genetically, and do not depend on a single feature which can be easily selected against. It is possible, though not easy, to raise dual cultures of plant cells and some pathogens, particularly fungi, in order to study host–pathogen interactions *in vitro*. Unfortunately, there are instances where such *in vitro* interactions do not correspond to observed *in vivo* effects. Some pathogen-resistant plants, for example, yield pathogen-sensitive tissue cultures. Similarly, pathogen-resistance mechanisms expressed in cultured cells may not always have a role in the resistance of the whole plant. This hinders attempts to use dual cultures for selection of resistant cells. Nevertheless, there has been some progress in the use of dual cultures to screen *in vitro* for pathogen-resistance traits which might otherwise be difficult to determine *in vivo*.

Some plant pathogens, however, damage tissue through the production of phytotoxins. This provides the opportunity of selecting cells with tolerance of the phytotoxin and regenerating disease-resisting plants. An example of this was the selection of *Nicotiana* cells with tolerance of methionine sulphoximine (MSO), a compound structurally similar to tabtoxin, a methionine antimetabolite produced by the 'wildfire' pathogen *Pseudomonas tabaci*. When plants were regenerated from the cells some were found to be resistant to tabtoxin and showed reduced disease symptoms when infected. Despite some discussion in the literature on the relationship between, and modes of action of, MSO and tabtoxin, this work illustrates how selection might be used to generate disease-resistant plants.

Similarly, *Solanum tuberosum* plants were regenerated from callus which had been selected in the presence of toxin-containing culture filtrates of *Phytophora infestans*. Leaves from the plants were more resistant to the toxin than controls, and the plants themselves had smaller lesions than normal after infection.

Recently, *in vitro* selection techniques were used successfully to produce cell cultures of *Citrus limon* with resistance to the toxin from *Phoma tracheiphila* (the fungal pathogen causing the 'mal secco' disease of lemon). The resistance was retained during transition of the cells from an undifferentiated callus state, through a differentiated embryo phase and back to callus again. Clearly, the next phase in such a selection programme will be to rigorously field-test the regenerated plants.

METAL TOLERANCE

Industrial activity and environmental pollution have led to the contamination of many areas with heavy metals such as lead, zinc, and cadmium. It has also been estimated that in as much as 40% of the world's arable soils the presence of aluminium may be inhibiting crop growth. This metal is a particular problem in areas with high acid rain because aluminium dissolves and mobilizes at low pH. Tolerances to metals such as these have been identified in higher plants, however, and cultured plant cells have been employed in order to provide an understanding of the mechanisms involved.

Datura innoxia cells, with tolerance of 250 μm cadmium were selected by incubating cultures in the presence of stepwise-increasing concentrations of cadmium chloride. The tolerance was retained for 400 generations in the absence of the metal and was attributed to production of metal-binding proteins by the tolerant cells. The proteins were not detected in unselected cadmium-sensitive cells but their synthesis was induced rapidly in the tolerant cells following cadmium treatment. It was considered unlikely, however, that the ability to produce the metal-binding proteins was completely absent in the sensitive cells. A more plausible explanation was that the tolerant cells were markedly over-producing proteins which were not normally required in such large amounts. The mode of tolerance, therefore, may have been similar in some respects to those involved in tolerances to other antimetabolites such as glyphosate and glufosinate.

Cultured cells and protoplasts of *Anthoxanthum odoratum* were used to study modes of tolerance to zinc and lead by examining effects of the metals on cell viabilities and respiration. It was shown that the cell wall was involved in lead tolerance because whereas intact tolerant cells were unaffected by the metal, their protoplasts were highly susceptible. Cells and protoplasts of non-tolerant *A. odoratum* were both susceptible to lead. In contrast, the cell wall did not appear to have a role in zinc tolerance because intact cells and protoplasts of tolerant clones were equally unaffected by the metal.

The solubility characteristics of aluminium present some interesting problems when studying the effects of the metal on cultured plant cells for selection work. At the pH of most plant cell culture media (approximately 5.8) aluminium precipitates and fails to inhibit growth unless present at excessively high concentrations. However, at pH 4.0, at which the metal is soluble, gelling of the agar in the medium is inhibited. In order to inhibit and select cells by aluminium under conditions which simulate those found in soils, therefore, it becomes necessary to modify the cell culture protocols. This has been achieved by growing the cells on filter papers supported by polyurethane foam filled with medium at pH 4.0. The medium also contained unchelated iron and lower phosphate and calcium concentrations in order to reduce aluminium precipitation. With adapted media such as this aluminium has been shown to cause toxic effects in cultures and the selection of aluminium-tolerant cells becomes plausible.

Genetic Engineering of Plant Cells

In the preceding sections we have seen how *in vitro* culture techniques can be used to select variant cells from which new plant varieties can be produced. Later we shall see how selection data have also been used to assist the design of experiments on the modification of plants using recombinant DNA (genetic engineering). It is evident from the discussion, however, that in some situations the potential of *in vitro* selection is limited by a number of difficulties. It is partly because of these difficulties that genetic engineering methods are assuming an increasingly important role in the modification of plants. We now consider some of these methods.

AGROBACTERIUM TUMEFACIENS–MEDIATED TRANSFORMATION

The plant pathogen *Agrobacterium tumefaciens* causes crown gall disease in higher plants. Infection by the bacterium commonly results in the formation of tumorous outgrowths, the morphology of which depends on the plant species. The bacterium is now known to possess a large (95–160 Md) tumour-inducing (Ti) plasmid, part of which (the T-DNA), is transferred and incorporated into the host genome during infection (Fig. 5.5). The tumours, which can be cultured *in vitro*, have been found to grow in the absence of phytohormones and produce a range of guanido amino acid compounds known as opines using T-DNA encoded enzymes. The specific opine type produced by a tumour depends on the infecting *A. tumefaciens* strain, but the commonest opines synthesized are octopine and nopaline. Demonstration of opine synthesis is often used as evidence for the presence of T-DNA in the cells.

Normally, *Agrobacterium*-transformed tumour tissues will not regenerate roots and shoots. However, mutants of *A. tumefaciens* with alterations in the T-DNA region of the Ti plasmid have now been found to produce tumours with altered morphology (Chapter 4). A knowledge of the effects of these plasmid genes on plant development has been important for the application of *Agrobacterium* to plant biotechnology because, by eliminating their activities from the plasmid to yield a 'disarmed vector', it has become possible to incorporate T-DNA into plant cell DNA while retaining the ability to regenerate roots and shoots. Specially designed Ti plasmid 'transformation vectors' have now been constructed which, typically, have the following characteristics:

- Absence of genes which interfere with normal plant differentiation
- A genetic marker such as the GUS gene and a selectable marker, for example conferring antibiotic resistance, to indicate the presence of the incorporated T-DNA
- Presence of the T-DNA 'border regions', i.e. the DNA regions which flank the left and right ends of the T-DNA and are required for the insertion of the T-DNA into the plant genome. (The significance of this is that foreign genes inserted into the T-DNA may also be incorporated into, and expressed in the host tissue to create new 'genetically engineered' plants.)

One of the problems with the manipulation of the Ti plasmid for use as a gene vector is its large size, which makes *in vitro* manipulation difficult. For example, a large plasmid is unlikely to have unique sites for cleavage by restriction enzymes in gene cloning experiments. This problem has been circumvented in two ways. In the first, Ti plasmids have been constructed containing a widely used gene cloning vector, such as pBR322, between the T-DNA border regions. Foreign genes

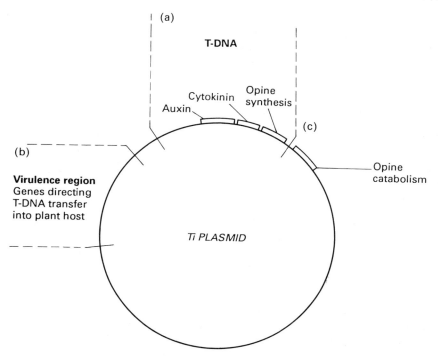

Fig. 5.5 A generalized Ti plasmid of *A. tumefaciens* showing: (A) The T-DNA region carrying genes coding for opine synthesis and auxin and cytokinin-like activities. Upon infection of a plant the T-DNA is transferred from the Ti plasmid into the plant cell genome. Foreign genes inserted into the T-DNA are, in this way, also incorporated into the plant DNA. (B) The virulence region coding for genes which bring about T-DNA transfer into the plant cell. (C) Genes for opine catabolism: *Agrobacterium* can utilize opines, synthesized in the plant under direction of the T-DNA, as nitrogen sources. The type of opine synthesized depends on the plasmid but the commonest are nopaline and octopine. Expression of the virulence region has been shown to be activated specifically by acetylsyringone and α-hydroxyacetosyringone. These compounds occur in exudates of wounded and metabolically active plant cells and may allow recognition of susceptible plant cells by the *Agrobacterium*.

carried on other pBR322 plasmids have then been co-integrated into the Ti plasmid by homologous recombination between the pBR322 sequences (Fig. 5.6). In the second method, binary plant vector systems have been employed (Fig. 5.7). These systems use two compatible plasmids, one of which is a broad host-range vector containing the T-DNA border regions flanking selective markers and other foreign genes. The second plasmid (a Ti plasmid, sometimes referred to as a helper plasmid) carries the virulence genes but has a large deletion in the T-DNA region.

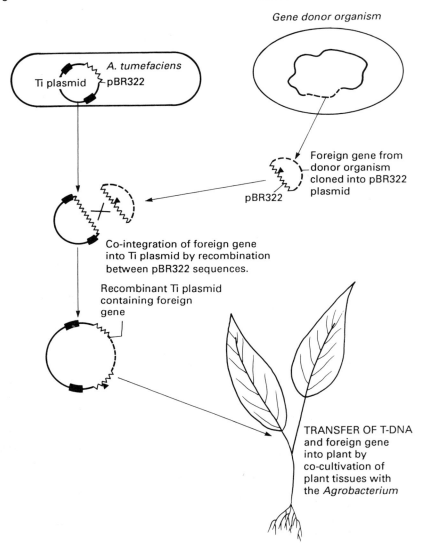

Fig. 5.6 Scheme for *A. tumefaciens*-mediated gene transfer into plants. This method uses a disarmed Ti plasmid (——) which contains a pBR322 copy (∼) between the T-DNA border sequences (■). Foreign genes (----) cloned into pBR322 plasmids (containing a selectable marker ▲) are co-integrated into the Ti plasmid by homologous recombination between the pBR322 sequences. Co-cultivation of plant tissues with *A. tumefaciens* containing the recombinant Ti plasmid then facilitates transfer of the DNA from between the T-DNA border sequences (including the foreign gene) into the plant genome. The plant tissues are freed of *A. tumefaciens* at a later stage by treatment with antibiotics.

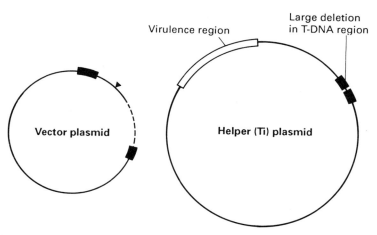

Fig. 5.7 Binary vectors for foreign gene transfer into plants. Binary vector systems employ two plasmids: a vector plasmid and a helper plasmid. The vector plasmid carries an inserted foreign gene (---) of interest and a selectable marker (▲) between its T-DNA border sequences (■). This plasmid has a broad host range and can replicate in *E. coli* and *A. tumefaciens*. The helper (Ti) plasmid has a large deletion in its T-DNA region but carries the virulence region required for T-DNA transfer, by complementation *in trans*, from the vector plasmid into the plant genome. The advantage of a binary system is that co-integration is not required (see Fig. 5.6). Transfer efficiencies of binary systems, however, may sometimes be lower than those of co-integration methods.

Both plasmids are transferred into *Agrobacterium* which is then used to infect plant tissues. Expression of the virulence region present in one of the plasmids then permits, by complementation *in trans*, transfer and integration of the T-DNA (including any foreign genes incorporated within it), from the second plasmid into the genomes of the host plant cells. The virulence genes themselves, however, are not stably transferred. The advantage of this method is that construction of a co-integrate is not required. This area of research is the subject of much academic and industrial activity and is progressing very rapidly. The procedures and vectors employed are being modified and improved constantly. Detailed accounts of methods and technical data, not provided here, may be found in other reviews (e.g. Lichtenstein and Draper, 1986). Early work on *Agrobacterium*-mediated transfer of foreign genes into plant cells used selectable antibiotic-resistance markers inserted into the T-DNA. In order to ensure expression of the integrated resistance marker in the host cell it was necessary to include transcription signals (promoter sequences) which were recognized by the plant transcription mechanisms. Typically this was done by inserting the new gene into the coding region for nopaline synthase which is expressed constitutively during tumour induction. This placed the antibiotic-resistance gene under the transcriptional control of the nopaline synthase promoter in the T-DNA causing its expression in the growing tumour. Plant cells transformed in this way were then able to grow in the presence of the antibiotic.

The potential of this area of research has been recognized especially by the world's leading agricultural chemical companies who are increasingly playing a major role in plant biotechnology and seed production. Experiments such as these helped in the development of procedures for the introduction of agriculturally important traits into plants. Some of the research on cell-selection for herbicide tolerances, for example, has now been extended through the use of *Agrobacterium* and recombinant DNA technology. The study of glyphosate tolerance illustrates this particularly well. It has not been possible by *in vitro* selection to isolate plant cells tolerant of glyphosate by virtue of their synthesis of a glyphosate-insensitive EPSP synthase enzyme. *In vitro* selected tolerant cells have usually been found to contain an overactivity of the normal glyphosate-sensitive enzyme and progress in regenerating tolerant plants from such cells has been slow (see earlier, p. 92). Researchers in the USA have circumvented this problem using two different approaches. In the first, a chimeric gene was constructed consisting of an EPSP synthase gene from *Petunia hybrida* connected to a promoter from CaMV. The promoter was known to result in a high level of expression of foreign genes in plants. The constructed gene was then introduced into *Agrobacterium tumefaciens* (by conjugational transfer) and transferred using the bacterium into leaf discs of normal plants. The leaf discs were then cultured to produce callus which displayed high levels of EPSP synthase activity (due to the presence of the virus promoter) and glyphosate tolerance. Later, plants were regenerated which were also found to be tolerant of the herbicide. The mechanism of tolerance in the genetically engineered plants (EPSP synthase overproduction through a high level of gene expression) mimicked that found in glyphosate tolerant cells isolated by *in vitro* selection (EPSP synthase overproduction through gene amplification) but the molecular mechanisms involved were different.

In the second approach, a mutated gene coding for an EPSP synthase enzyme with reduced sensitivity to glyphosate was isolated from *Salmonella typhimurium*, fused to plant regulatory signals, and transferred, using *Agrobacterium*, into petunia and tobacco plants. The transformed plants were then found to synthesize the bacterial enzyme and grew after spray-treatment with the herbicide. The expression of a bacterial gene to provide herbicide tolerance in a plant is a significant advance in plant biotechnology and demonstrates clearly the potential of this area of research.

Earlier, we discussed how tolerance of the herbicide glufosinate in selected cultured cells was attributed to overproduction of the target enzyme, glutamine synthetase, through gene amplification. Recently, tolerance of the herbicide was engineered into *N. tabacum* plants through the introduction of a detoxifying enzyme. The bacterium *Streptomyces hygroscopicus* contains a gene ('*bar*') coding for an enzyme which converts glufosinate into a non-toxic acetylated form. The '*bar*' gene was isolated, placed under the control of a CaMV promoter and transferred into a recombinant *Agrobacterium* plasmid. The bacterial gene was then in-corporated into plant cell genomes by co-cultivating the *Agrobacterium* with *N.*

tabacum leaf protoplasts. Calluses grown from the transformed protoplasts were found to express the '*bar*' gene and display resistance to glufosinate. Moreover, when whole plants were produced containing the transferred gene they also showed marked resistance to the herbicide even when applied at rates above those normally used in agriculture.

Genetic engineering is also now being used to develop new plant varieties with reduced symptom expression during viral infection. Some cucumber mosaic viruses contain a small satellite RNA molecule which is dependent on the virus for replication. The satellite RNA molecule is, therefore, rather like a parasite of the virus. The consequence of this is that symptoms produced by the virus containing the satellite RNA are (for reasons that are not understood) much reduced compared with the effects of satellite-free virus on plants. Presence of the satellite RNA may also lead to a reduced number of viruses in the infected plant. It was postulated that incorporation of the satellite RNA by genetic engineering into plants would lower their susceptibility to the viral disease. The RNA molecule was, therefore, copied into a DNA molecule and transferred into *Agrobacterium* on a vector derived from the *Agrobacterium* Ti plasmid. The vector also contained a gene coding for resistance to the antibiotic kanamycin in order to aid selection of the transformed cells. The *Agrobacterium* was then used to infect cells of *N. tabacum* from which kanamycin-resistant plants were later regenerated, so demonstrating the presence of the newly inserted genes. More importantly, however, infection of the plants with cucumber mosaic virus caused replication of satellite RNA from the inserted DNA template which then parasitized the virus and so reduced the symptom development. This is one of the first examples where genetic engineering has been used to reduce the susceptibility of a plant to disease. In a similar series of experiments, the gene for the coat protein of tobacco mosaic virus (TMV) was inserted into the tobacco genome using an *Agrobacterium* Ti plasmid-based vector. Regenerated transgenic plants, which expressed the viral coat protein, showed delayed symptoms in response to TMV infection. Although one phenomenon is not well understood, this type of resistance is thought to be related to the 'acquired resistance' shown by many plants following viral infection, one resistance usually being restricted to viruses related to the primary infecting virus. The result of this experiment clearly implicates the viral coat protein as the determinant of resistance in this case.

The types of foreign genes which can be inserted by *Agrobacterium* into plant cells are not limited to those from other plants or microorganisms. It has now proved possible to incorporate the gene for human growth hormone (hGH) into tobacco and sunflower cells. Normally in human cells the hGH gene is transcribed to produce a pre-mRNA containing 'introns' (non-coding sequences) which must be removed by processing in the cell nucleus to produce functional mRNA. Early work on the transfer of the hGH gene into plants, however, showed that although the gene was transcribed into pre-mRNA it was not processed correctly. Clearly, therefore, the processing signals on the hGH gene were not recognized by the plant cells. Such experiments provide interesting information on the differences between pre-mRNA processing in plants and animals and it is conceivable that a future biotechnological application may be found. More recently, the genes encoding mouse IgG and human serum albumin have been cloned into tobacco plants and apparently expressed faithfully. The prospect of utilising crop plants as a source of pharmaceutical proteins is clearly an exciting one, and may present advantages compared with expression of such genes in microbes and yeasts.

Much of the research on *Agrobacterium* to date has been carried out using dicotyledonous plants. This is mostly because monocotyledonous species, which are generally difficult to culture *in vitro*, appeared to be insensitive to *Agrobacterium*. This is unfortunate because many of the most important agricultural crops are graminaceous monocotyledonous species. Since the mid 1980s, however, the situation has begun to change with reports describing the uptake and expression of T-DNA genes in a number of monocotyledonous plants. Two species, *Chlorophytum capense* (Liliaceae) and *Narcissus* cv. Paperwhite (Amaryllidaceae), were shown to express the genes for octopine and nopaline synthase after infection with *Agrobacterium* but did not produce large crown gall tumours; the wound sites displayed only slight swelling. However, cultured stem fragments of another monocotyledon, *Asparagus officianalis*, did produce tumour cells which grew *in vitro* without phytohormones and synthesized opines for several cell generations. Some progress has also now been made in transforming *Zea mays* with *A. tumefaciens* and it seems likely that further progress will follow.

Recently, a greater interest has been developing in an alternative *Agrobacterium* species, *Agrobacterium rhizogenes*, for the genetic manipulation of plant cells. This bacterium causes 'hairy root' disease in plants through alterations in auxin metabolism after transfer of genes from a large root-inducing Ri plasmid into the plant genome. Plants or callus cultures infected with the bacterium produce a proliferation of 'hairy roots' which grow in the absence of phytohormones. The effect is analogous to that produced by the Ti plasmid of *A. tumefaciens* and the Ri plasmid is, potentially, a good vector for foreign genes. The virulence genes of the Ri plasmid will also promote the integration of T-DNA from Ti plasmid and hence can be used with Ti-derived binary vectors. The effects of *A. rhizogenes* have now been studied in a number of economically important plants especially *Brassica napus* (oilseed rape). One of the features of 'hairy roots' is that they are capable of regenerating into whole plants. The phenotypes of the regenerants, however, are altered and show wrinkled leaves and reduced apical dominance. The plants can also be shown to contain Ri T-DNA, synthesize opines and transfer the phenotype to their progeny. A further feature of 'hairy root' cultures is that they can be induced to form undifferentiated suspension cultures by the addition of phyto-hormones. Removal of the phytohormones then allows the cells to reorganize into hairy roots. This mechanism presents the possibility of selecting cells for desirable traits in an undifferentiated state from which organized hairy root cultures and, eventually, whole plants can be regenerated. Our understanding of *A. rhizogenes* has in the past followed our understanding of *A. tumefaciens*, and it seems likely that the Ri plasmid will assume greater importance in plant gene manipulation in the future. Other aspects of research on *A. rhizogenes* concerning secondary metabolite production are discussed elsewhere in this volume (see Chapter 6).

DIRECT GENE TRANSFER INTO PLANTS

As a consequence of the limited host range of *Agrobacterium*, alternative gene transfer systems have been sought. One possibility, which has been tried with some success, is the direct transfer of genes into protoplasts. In the absence of the cell wall, uptake of naked DNA into the protoplast becomes feasible. Gene constructions have been prepared consisting of marker genes (conferring resistance to antibiotics) and gene expression sequences from e.g. the nopaline and octopine synthase genes or gene VI of the cauliflower mosaic virus. Protoplasts are incubated with the DNA at 22–24 °C for about 30 min in the presence of PEG. Alternatively, protoplasts are subjected to electroporation in the presence of the DNA to be inserted (see Chapter 3). During these treatments the recombinant DNA construct is taken up into the protoplasts which are then washed and cultured. In this way it has been possible to transfer genes directly into protoplasts of graminaceous plants such as *Triticum monococcum* and *Lolium multiflorum*. The presence of the genes in the transformed cells can be determined by assessing levels of antibiotic resistance. Such progress has shown that the genetic engineering of monocotyledons is a viable proposition. One of the major remaining problems with cereals, however, is our lack of knowledge of reliable systems to regenerate whole plants from culture. Similar experiments carried out with dicotyledons such as *N. tabacum* have shown that DNA transferred into cells by direct uptake can be incorporated into the plant chromosomal DNA and transferred to sexual progeny. Usually the inheritance of the inserted gene follows Mendelian rules. It seems likely that the same will apply to monocotyledons if problems with plant regeneration from protoplasts can be overcome, in which case the genetic manipulation of cereal crops will advance much more rapidly.

Another method of inserting genes directly into plant cells is by microinjection. This work requires the provision of some finely adjustable equipment for the micromanipulation and injection of cells. In Japan, researchers have demonstrated the microinjection of berberine-labelled DNA into protoplasts. Berberine fluoresces yellow under ultraviolet light and was easily visualized in the cells using a microscope fitted with epifluorescence optics. Protoplasts of *Euphorbia millii* were held by a miniature holding pipette under gentle suction and injected with the DNA using a pressurized glass micropipette. The injected protoplasts remained viable. With careful orientation of the protoplasts it was also possible to specifically inject the cell nucleus with the labelled DNA.

VIRUSES AS GENE VECTORS

CaMV has also been employed as a vector for gene transfer into plants. The virus has a relatively small, double-stranded DNA molecule with some unusual single-stranded interruptions. CaMV has a number of features which recommend its use as a vector. The small genome is relatively easy to modify *in vitro* and the DNA is infectious when applied directly to the leaves. The infection then spreads rapidly throughout the plant and virus particles can be found in most of the cells. Parts of the CaMV genome appear to be non-essential for infection and can be deleted or replaced with foreign DNA. An example of this was the insertion of an *Escherichia coli* gene (*dhfr*) coding for dihydrofolate reductase, which confers resistance to methotrexate. When the CaMV DNA containing the *E. coli* gene was used to infect *Brassica rapa* (turnip) it was found that the *dhfr* gene was expressed in the plants and a functional dihydrofolate reductase was produced. The virus-infected plants were also little affected by spray treatments of methotrexate. Clearly therefore, CaMV has potential for the transfer of agriculturally desirable traits into crops.

There are, however, many properties of plant viruses which limit their potential for use as gene vectors. There are constraints on the size of the foreign DNA which can be inserted into the viral genome, without removing essential viral genes for replication. The upper limit for CaMV, for example, is thought to be less than 1 kbp of foreign DNA, or less than the size of an average eucaryotic gene. Attempts are being made to circumvent this problem through the use of binary vectors, i.e. one viral genome carrying foreign genes, and the second carrying viral genes essential for replication, etc. However, there are also general problems of disease symptoms which may be produced after expression of viral genes in the host plant. Clearly, much more research will be required before viruses acquire a major role in plant genetic engineering.

Conclusion

The improvement of plants via cell culture and genetic engineering techniques is an area of research which is progressing at an ever-increasing pace. In a short review such as this it is not possible to cover in depth every facet of the subject, and the reader is referred to the book by Walden in this series. The topics chosen for discussion serve merely to illustrate some of the activities which are taking place. Most of the manipulated plants described here have not yet been applied to commercial production. Nevertheless, the techniques present some exciting possibilities for agriculture and the extent of future advances is difficult to predict. As with all biotechnology, however, we must temper our enthusiasm with a regard for the consequences of our activities on the environment. This will undoubtedly form the basis of much discussion in the future, and the practical and commercial application of techniques which are clearly here to stay will require a great deal of thought.

References and Further Reading

Chaleff, R.S. (1981). *Genetics of Higher Plants. Applications of Cell Culture*. Cambridge University Press, Cambridge.

Chaleff, R.S. and Parsons, M.F. (1978). Direct selection *in vitro* for herbicide-resistant mutants of *Nicotiana tabacum*. *Proceedings of the National Academy of Sciences, USA* **75**, 5104–5107.

Dix, P.J. and Street, H.E. (1976). Selection of plant cell lines with enhanced chilling resistance. *Annals of Botany* **40**, 903–910.

King, P.J. (1984). From single cells to mutant plants. *Oxford Surveys of Plant Molecular and Cellular Biology* **1**, 7–32.

Klee, H., Horsch, R. and Rogers, S. (1987). *Agrobacterium*-mediated plant transformation and its further applications to plant biology. *Annual Review of Plant Physiology* **38**, 467–486.

Lichtenstein, C. and Draper, J. (1986). Genetic engineering of plants. In *DNA Cloning*, Vol. 2 (D.M. Glover, ed.), IRL Press, Oxford, 67–117.

Maliga, P. (1984). Isolation and characterisation of mutants in plant cell culture. *Annual Review of Plant Physiology* **35**, 519–542.

Meins, F. (1983). Heritable variation in plant cell culture. *Annual Review of Plant Physiology* **34**, 327–346.

Radin, D.N. and Carlson, P.S. (1978). Herbicide-tolerant tobacco mutants selected *in situ* and recovered via regeneration from cell culture. *Genetics Research* **32**, 85–89.

Shaner, D.L. and Anderson, P.C. (1985). Mechanism of action of the imidazolines and cell culture selection of tolerant maize. In *Biotechnology in Plant Science. Relevance to Agriculture in the Eighties*, Academic Press, London, pp. 287–299.

Vasil, I.K., Scowcroft, W.R., and Frey, K.J. (eds) (1982). *Plant Improvement and Somatic Cell Genetics*. Academic Press, London.

Chapter 6

Natural Products and Metabolites from Plants and Plant Tissue Cultures

ANGELA STAFFORD

The Importance of Plant Products

The use of plants as a major ingredient of the human diet is a subject beyond the scope of this book. Also important, however, is the role that plant metabolites have played through the centuries as medicinal agents, mainly in the form of crude extracts or 'herbal' cures. Certain of these remedies have assumed a key position in conventional medicine. In addition, plant pigments and perfumes have achieved considerable importance in the cosmetic and food industries. It is on these often relatively minor constituents of plants, mostly so-called 'secondary metabolites', that this discussion will concentrate, with particular emphasis on the role that plant cell and tissue cultures can play in their production.

EVIDENCE FROM FOLKLORE

For many centuries prior to the development of phytochemical extraction and analytical methods, herbal remedies played a major part in the treatment of disease in Western civilizations. Indeed, in countless societies throughout the world today, plants feature significantly in local cures for common ailments. It has often been argued that the traditional use of a natural remedy provides evidence for a real biological activity, and some screening programmes do in fact take such information into consideration. Some notable successes have resulted from this type of strategy. For example, recent phytochemical analysis of plants used in

Table 6.1 Application of plant products in industry

Product	Application	Plant source
Medicinals		
Codeine	Analgesic	*Papaver somniferum*
Atropine	Anticholinergic	*Atropa belladonna*
Digoxin	Cardiatonic	*Digitalis lanata*
Quinine	Antimalarial	*Cinchona ledgeriana*
Vincristine	Antileukaemic	*Catharanthus roseus*
Food flavours or additives		
Quinine	Bittering agent	*Cinchona ledgeriana*
Thaumatin	Sweetener	*Thaumatococcus danielli*
Lycopene	Red pigment	*Lycopersicon esculentum*
Crocin	Yellow pigment	*Crocus sativus*
Nicotine	Insecticide	*Nicotiana tabacum*
Pyrethrin	Insecticide	*Chrysanthemum cinerariaefolium*
Jasmine oil	Perfume	*Jasminum* sp.

folklore for the treatment of cancer has yielded a number of compounds with antitumour activity. Among these are usnic acid, derived from lichens, and podophyllotoxin from *Podophyllum hexandrum* and related species, a plant used by the ancient Chinese as an antitumour drug. Reference to folklore was also responsible for the isolation of the important alkaloidal anticancer drugs from the Madagascan periwinkle (*Catharanthus roseus*), although in this case the traditional use of the plant had been for the treatment of diabetes.

In those circumstances in which the rewards are likely to be great, random screening can also be employed. Of increasing interest to the large pharmaceutical firms are new antiviral compounds. Plants of a number of families have recently been shown to accumulate a class of alkaloids with anti-HIV activity, notably castanospermine from the Australian Morten Bay chestnut tree. There is no doubt that the biosynthetic repertoire of the plant kingdom is enormous, and that many of the pharmacologically active compounds produced are unique, being amenable to chemical synthesis only with great difficulty. The discovery of new, bioactive natural products also supplies the chemist with skeletons upon which to build totally novel compounds with enhanced activities.

PLANT METABOLITES IN INDUSTRY

As already mentioned, plant products have a wide range of applications in several industries. Table 6.1 lists just a few examples with a relatively widespread use, and their chemical formulae are given in Fig. 6.1. Codeine, atropine, and digoxin are

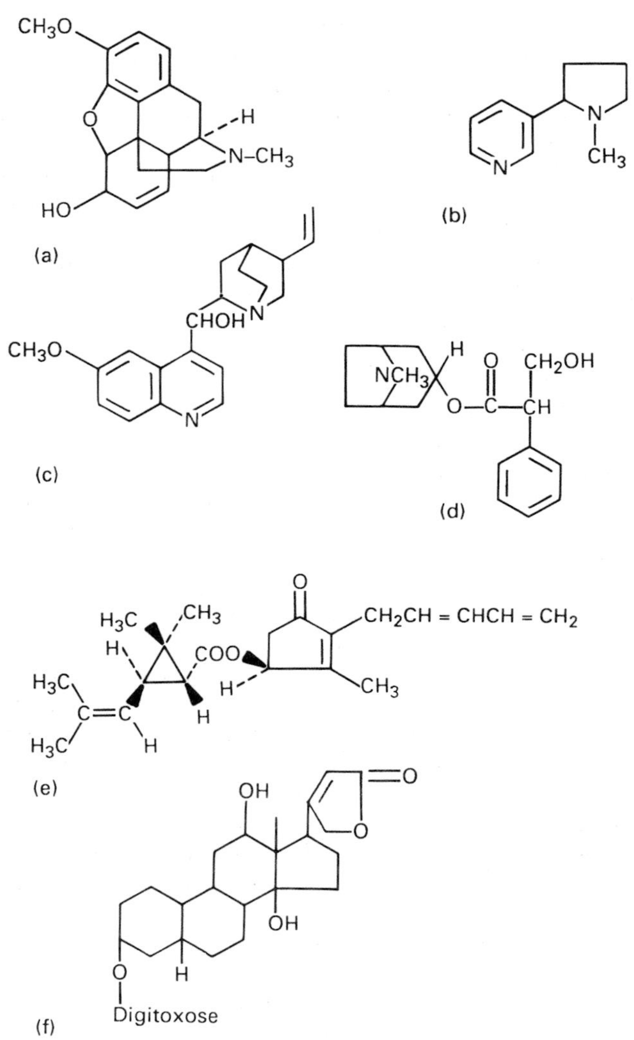

Fig. 6.1 Industrial plant products (see Table 6.1). (a) Codeine, (b) nicotine, (c) quinine, (d) atropine [(−)-hyoscyamine], (e) pyrethrin I, (f) digoxin.

three of the most frequently prescribed drugs derived from plants. The alkaloid codeine is extracted from crude opium along with its biosynthetic derivative morphine, has milder sedative properties than the latter, and is particularly useful in alleviating coughing. Atropine is one of a group of so-called 'tropane alkaloids' which are accumulated by many members of the potato family, including *Atropa belladonna* (deadly nightshade). Atropine is anticholinergic, and derivatives are used in pre-anaesthetic medication, one property of the drug being to suppress secretions. Digoxin is one of a number of related cardioactive steroids found in *Digitalis* (foxglove) species. Quinine is the most important alkaloid to be found in the bark of the *Cinchona* tree. In one species, *C. ledgeriana*, the alkaloid reaches levels of about 8%, and this tends to be the source of quinine not only for antimalarial preparations, but also as an additive in the food industry. Quinine is one of the most bitter substances known and it is this property which makes it such an important constituent of tonic water. Quinidine, a related alkaloid, is also found in *Cinchona* bark but has a totally different usage as a cardiac depressant.

Vincristine and vinblastine are both antineoplastic alkaloids extracted from the leaves of *Catharanthus roseus*, where they are present in very low levels (less than 0.001%). The demand for these compounds and their high market value has stimulated numerous efforts to produce them via plant cell culture. Progress made over the last couple of years has been particularly interesting and this will be discussed in a later section (pp. 154–60).

Thaumatin has excited considerable interest because of its potential value as a low-calorie sweetener. It consists of two proteins, found in the underground fruits of an African plant, *Thaumatococcus danielli*. Its taste is similar to that of liquorice, but it is said to be about 2000 times sweeter than sucrose. Two other intensely sweet proteins, monellin and miraculin, are also found in African plants; *Dioscoreophyllum cumminsii* and *Synsepalum dulcificum*.

Lycopene and crocin are both carotenoid pigments currently experiencing an increase in favour with the food industry, as the synthetic red and yellow food colourants have become less popular. Lycopene is a red pigment found in a number of ripe fruits including the tomato, from which about 20 mg lycopene can be extracted per kilogram of fresh tissue. Crocin is the yellow-orange constituent of saffron, a name given to the stigmas of *Crocus sativa*.

Nicotine and pyrethrin are both highly effective insecticides. While pyrethrin tends to be applied as a liquid or solution, nicotine is often an ingredient of fumigants. Both substances are also toxic in humans, nicotine being readily absorbed through the skin.

While all the substances mentioned above are single chemicals, most preparations used by the perfume industry are mixtures of essential oils. In jasmine, for example, the main odour component is jasmone, though the total aroma is mediated by many other volatiles, mainly lower terpenoids.

Plant enzymes presently have a fairly limited usage in industry, microbial fermentations providing an alternative and generally less costly source. The most important plant enzymes used in the food industry are β-amylase and papain. β-Amylase is produced traditionally in germinating barley grains prior to the formation of the 'mash' in brewing. Papain is a thiol proteinase derived from the pawpaw (*Carica papaya*) where it is accumulated to high concentrations. Other plant proteinases with similar activities are bromelain from pineapple and ficin from fig. Papain and related enzymes from pawpaw have several applications in the food industry, amongst which are the prevention of 'haze' production in beer, and the tenderizing of meat. In 1988 the world market for this enzyme was estimated at around $22 million.

THE DISCOVERY OF NEW PLANT PRODUCTS CONTINUES

The search for new and useful plant products continues, with most emphasis being placed upon those with novel or enhanced pharmacological properties. Screening programmes aimed at the new antiviral, anticancer, and antimalarial compounds have over the last few years been successful in the discovery of many potentially exciting phytochemicals.

A particularly interesting story relates to the development of antimalarial drugs. Despite efforts to control the *Anopheles* mosquito, an estimated 215 million people are chronically affected by the disease, and there are still 150 million new cases every year. Quinine has long been used as the major drug in the treatment of malaria and is currently undergoing a revival in its popularity. However, in 1979, the structure of a new plant drug artemisinin (Fig. 6.2) was determined. This terpenoid compound was isolated from *Artemisia annua*, which is used as a traditional Chinese remedy for the treatment of malaria. In one trial, the treatment of cerebral malaria patients with artemisinin resulted in over 90% cured cases. This result was especially significant as the study was carried out in an area in which the malarial parasite *Plasmodium* was highly resistant to conventional drug therapy. Not surprisingly, this finding has stimulated still further effort into the identification of novel antimalarial drugs. Other plant products exhibiting high levels of activity in antimalarial screens include quassinoids, for example bruceantin (see Fig. 6.2) which also has potent anticancer properties.

Recognition of the involvement of HIV in AIDS has led to a surge of interest in the production of new antiviral compounds which might be employed to treat the disease. A group of scientists working at the Royal Botanic Gardens, Kew, have

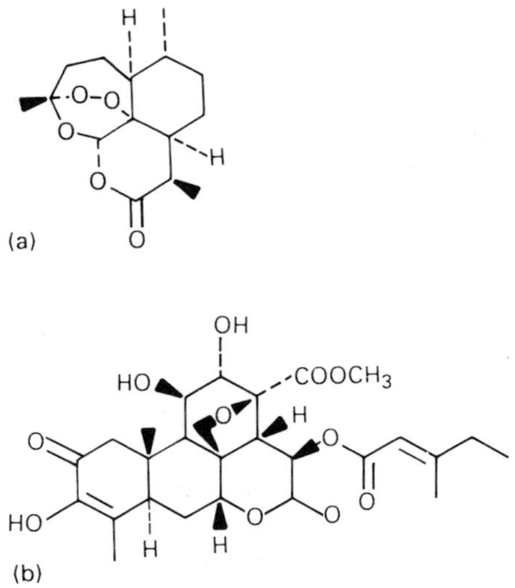

Fig. 6.2 Antimalarial compounds. (a) Artemesinin, (b) bruceantin.

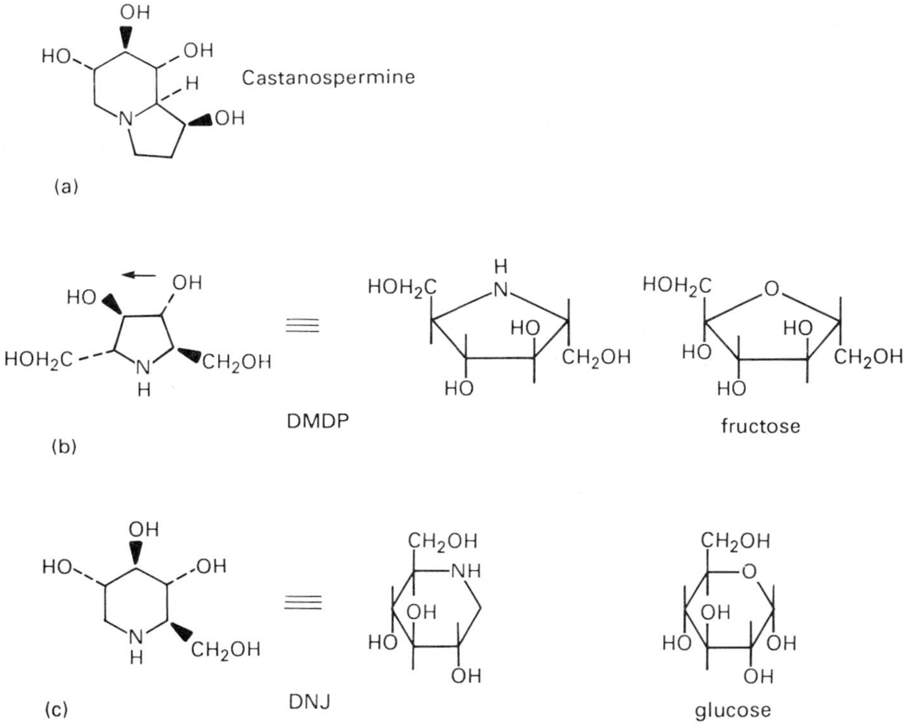

Fig. 6.3 Anti-HIV compounds from plants. (a), Castanospermine, (b) DMDP, (c) DNJ.

recently been conducting a screening programme for phytochemicals possessing the ability to inhibit glycosidase enzymes. Their main objective was the isolation of insecticidal compounds with this biochemical activity. However, in antiviral screens it was found that replication of the AIDS virus, which has a heavily glycosylated coat protein, was inhibited by a number of these compounds. Among these are castanospermine from *Castanospermum australe*, DMDP from *Lonchocarpus* spp., and DNJ (desoxynojirimycin) from *Morus* spp. (mulberries), the structures of which are shown in Fig. 6.3. One fascinating aspect of these chemicals is their similarity in structure to sugars; hence their specific enzyme-inhibitory properties.

These two cases exemplify how folklore, effective screening programmes, and an element of good fortune are turning up previously unexplored natural products of immense importance. The recent revival of interest in the intrinsic value of the world's flora as a source of novel pharmaceuticals is resulting in a concern that this great natural resource is in danger of becoming severely reduced, for example, via the destruction of the South American rain forests, vast areas of which are obliterated each year.

Biosynthesis and Accumulation of Plant Secondary Products

In the remainder of this chapter the discussion will concentrate on plant secondary metabolites only. Over the years, many definitions of a 'secondary metabolite' have been proposed, and probably the most acceptable simple interpretation is of any chemical which is not essential for the survival of the organism. In fact in many cases the opposite is true, in that secondary products are often potentially positively detrimental to the metabolism of the producer plant or microbe. The toxic nature of such chemicals is a feature which has been selected for in providing many plants with defence systems, against insect, fungal, or herbivorous attack, and many species have evolved highly sophisticated compartments for the accumulation of cytotoxic substances, such as glandular hairs, or laticifers. For many of the more water-soluble secondary products, the cell vacuole provides the major site of accumulation.

METHODS OF DETECTION

Some types of secondary metabolites can be visualized in plant cells by the use of light microscopy. These include tannins, anthocyanins, and carotenoids as well as some compounds which fluoresce when irradiated with certain wavelengths of ultraviolet light, such as the alkaloid serpentine found in *Catharanthus roseus*; this can be located in the cell vacuole as an electric-blue fluorescence when excited at around 360 nm.

However, the vast majority of important plant secondary compounds are colourless and not detectable by such convenient methods. Instead, for detection we must depend upon a combination of good chemical extraction procedures and the best available separation methods, preferably followed by purification and confirmation using critical chemical and physical procedures. When dealing with cell or tissue cultures rather than the whole plant tissue from which a target compound is normally isolated, several factors should be considered.

To begin with, extraction methods designed for whole plants cannot always be extrapolated to the cell culture. There are several reasons for this, the most important of these being the fact that in many cases, cell cultures only accumulate a fraction of the levels of secondary products found in the source plant. To produce an extract in which the desired compounds are detectable by standard methods, it is therefore often necessary to concentrate down much more than usual. This strategy results in a relative abundance of contaminating 'unknowns' unless the extraction procedure is extremely discriminating. A second problem is that, especially in undifferentiated tissue cultures, the possibility exists that secondary metabolites may be stored in a different compartment to that found in the whole plant, or bound in such a way that they cannot be retrieved by conventional extraction methods. A general text on phytochemical techniques is Harborne (1973).

Having determined a suitable extraction method, the next step is usually to make a preliminary identification of target compounds using one of a number of different separation methods. Amongst these are HPLC (high-performance liquid

Table 6.2 Immunoassays of plant metabolites in plant tissues

Species	Target compound	Type of assay	Sensitivity (pg)
Digitalis lanata	Digoxin	Radioimmunoassay	100
Datura sanguinea	Scopolamine	Radioimmunoassay	200
Catharanthus roseus	Ajmalicine Serpentine	Radioimmunoassay	100 (ajmalicine) 500 (serpentine)
Lycopersicon esculentum	Zeatin riboside	Radioimmunoassay	20
Pisum sativum	Abscisic acid	ELISA	20
Catharanthus roseus	Vindoline	ELISA	5
Picrasma excelsa	Quassin	ELISA	5
Cinchona ledgeriana	Quinine Quinidine	ELISA	10
Atropa belladonna	Tropic acids (e.g. atropine)	ELISA	100

chromatography), GC (gas chromatography), paper electrophoresis, or TLC (thin-layer chromatography). In each of these cases the sample extract is compared under the same running conditions with a pure standard. Those compounds appearing in the extract and exhibiting the same characteristics as the standard may be collected for further investigation. Some features which may be considered in this comparison with the standard are retention time on the separation matrix (e.g. silica gel, ion exchange resin, paper), as well as spectral characteristics in the case of HPLC or staining reaction with a coloured reagent. An alternative way of separating specific molecules from a crude mixture is to make use of antisera raised against the compound in question. Polyclonal or monoclonal antibody preparations can be used very effectively to detect plant metabolites, as has been shown for a large number of examples including quinine, digoxin, morphine, vinblastine, and scopolamine. These radioimmunoassays or enzyme immunoassays (depending upon the label) are now routinely used for screening of drug levels in blood but have also been adapted specifically for their application to plant extracts (Table 6.2). The production of immune sera for specific molecules also means that in theory at least, localization of these in plant tissues at the cellular and subcellular level should be possible, using such techniques as gold-immune labelling.

Two or more of these detection methods can of course be used in conjunction to offer a more positive identification, and they can also be used in a preparative mode to collect 'possibles' for more definitive analysis by e.g. NMR (nuclear magnetic resonance), IRS (infrared spectroscopy), or MS (mass spectrometry).

The plant organs or tissues within which secondary products are accumulated are not necessarily their sites of synthesis. A number of techniques have been developed to investigate these diverse pathways and the sites of biosynthesis, including grafting (e.g. of non-alkaloid producing plant aerial organs on to the roots of alkaloid-producers), and the treatment of sterile tissue cultures or excised organs with radioactive precursors. Localization of key secondary biosynthetic enzymes has in some cases furthered our knowledge of the ways in which these complex metabolic pathways can be achieved and compartmentalized. We will explore the use of these methods using the examples first of the tobacco alkaloids, and second of the morphinane alkaloids found in *Papaver* species.

The main species used commercially for the production of tobacco is *Nicotiana tabacum*, but much of the extensive work on the biosynthesis of the *Nicotiana* alkaloids has been carried out using other species in the same family, e.g. *N. rustica* and *N. glauca*. Nicotine and related alkaloids are classified as pyridine alkaloids. Although nicotine is the main alkaloid in nearly all species of *Nicotiana*, more than 40 other pyridine alkaloids have been identified, including nornicotine, anabasine, and anatabine (Fig. 6.4). Relative proportions of these four alkaloids in *N. tabacum* were found to be nicotine 93%, anatabine 3.9%, nornicotine 2.4%, and anabasine 0.35%.

A widely adopted approach to the investigation of alkaloid biosynthesis in *Nicotiana* species has been the administration of radiolabelled compounds. Many

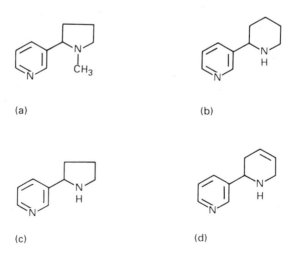

(a)   (b)

(c)   (d)

Fig. 6.4 Pyridine alkaloids of *Nicotiana* sp. (a) Nicotine, (b) anabasine, (c) nornicotine, (d) anatabine.

Table 6.3 Tracer experiments with *Nicotiana glauca* and [14]C- or ^{3}H-labelled precursors

Labelled precursor compounds	Alkaloids labelled	Incorporation (%)
Using whole plants and the 'cotton wick' method		
[2-^{14}C] Acetate, sodium	Anabasine	0.1–2.0
DL-[2-^{14}C] Lysine	Anabasine	4.0–6.4
[2-^{14}C] Nicotinic acid	Anabasine	14.3
	Anatabine	0.2
	Nicotine	0.1
	Nornicotine	0.1
Using excised shoots		
[2-^{14}C] Acetate, sodium	Anabasine	0.03
[1,5-^{14}C] Cadaverine	Anabasine	0.49
DL-[2-^{14}C] Lysine	Anabasine	0.04
(*RS*) [2′-^{14}C] Nicotine	Nicotine	0.41
	Nornicotine	5.70
	Anabasine	0
Using excised roots		
L-[U-^{14}C] Lysine	Anabasine	28.6
[^{3}H] Nicotinic acid	Anabasine	12.0

different methods of feeding have been used, including hydroponics, in which the isotope is added to the nutrients in which the roots of the intact plant are growing, and wick feeding, in which a cotton wick is inserted into the plant stem and placed into contact with a solution of the isotope. In addition, excised organs have often been exploited to gain information on the site of synthesis within the whole plant. Cell cultures have also been used to examine biosynthetic routes in view of their large surface area which is presented to the external medium.

Some data resulting from such experiments are shown in Table 6.3, and bearing in mind the fact that in *N. glauca* anabasine is the major alkaloid, a number of conclusions can be drawn:

- Lysine is the source of the alkaloid pyridine ring in *Nicotiana*
- Nicotinic acid is a direct precursor of anabasine and related alkaloids
- Roots are the major site of synthesis of anabasine from lysine
- Shoot tissues are able to demethylate nicotine to nornicotine.

Similar experiments have resulted in the elucidation of the tobacco alkaloid biosynthetic pathway, shown in Fig. 6.5. Grafting experiments, whereby *N. tabacum* shoots were grafted on to roots of other members of the family *Solanaceae*, e.g. potato, and vice versa, have also confirmed that these alkaloids are formed predominantly in the roots of the plant. These observations have led to the investigation of differentiated root cultures as production systems for *Nicotiana*

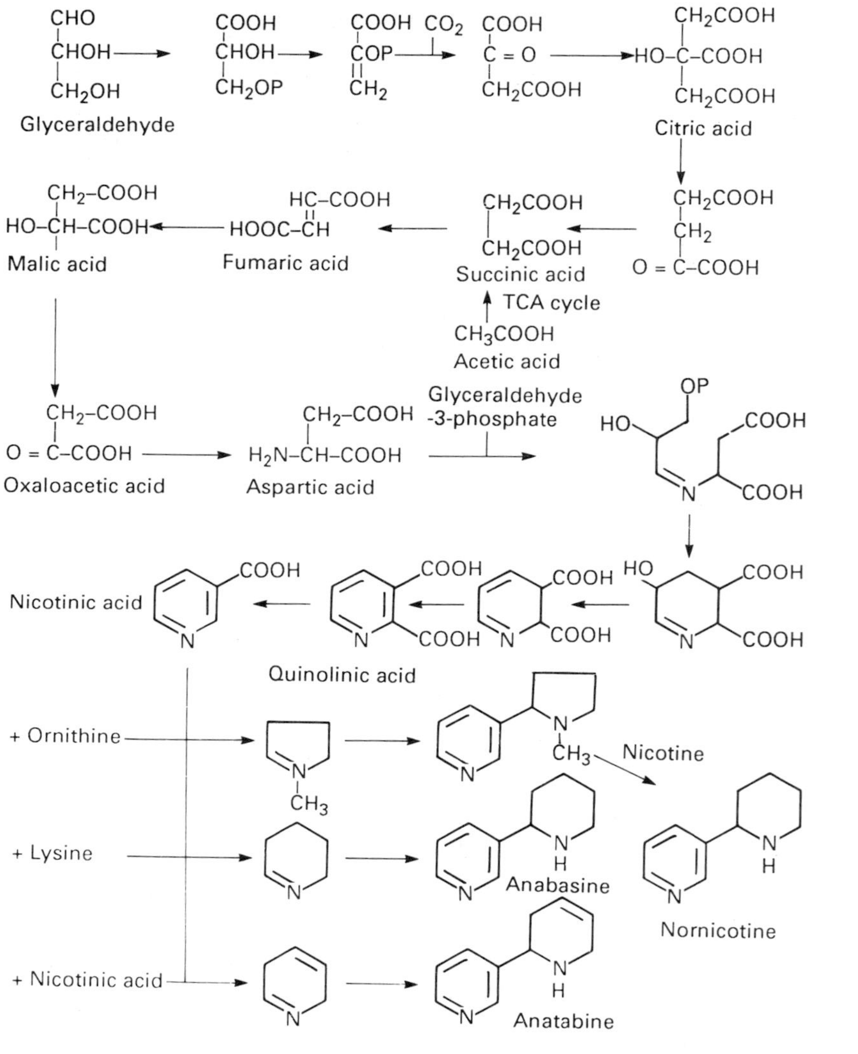

Fig. 6.5 Biosynthesis of tobacco alkaloids.

alkaloids. The more medicinally important alkaloids, hyoscine and hyoscyamine, produced by other members of the same family, for example *Atropa belladonna* and *Datura* species, can also be produced in this way, and the procedure will be discussed further on pp. 149–50.

Members of the *Papaveraceae*, including *Papaver somniferum*, *P. bracteatum*, and *Macleaya cordata* accumulate a number of alkaloids deriving from phenylalanine via dopamine. The main site of accumulation of these compounds in the whole plant is laticifers, specialized cells arising in the seedlings a few days after

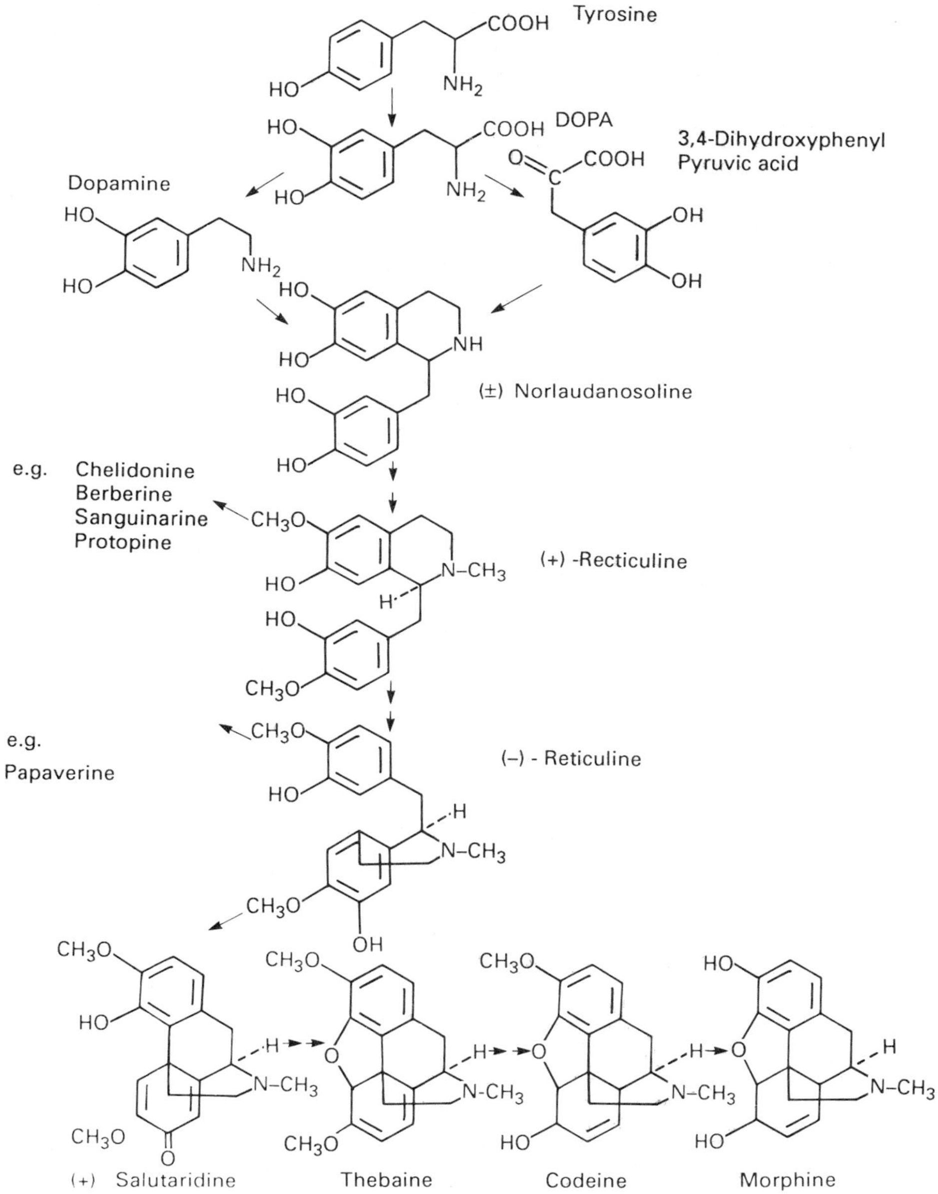

Fig. 6.6 Biosynthesis of some *Papaver* alkaloids.

Table 6.4 Tissue culture systems investigated for the production of morphine alkaloids

Species	Culture type	Alkaloids detected
Papaver bracteatum	Callus	Thebaine (60 µg/g dry weight)
	Callus with shoots	Thebaine (70 µg/g dry weight)
	Suspension	Thebaine (no quantitation made)
Papaver somniferum	Callus	Thebaine (1 mg/g dry weight)
		Codeine (33 µg/g dry weight)
	Embryoids	Thebaine 2 mg/g dry weight)
	Suspension	Codeine (1.5 mg/g dry weight)

germination which are distributed throughout the entire plant. The laticifers in *Papaver* species are articulated and branching, giving rise to networks of latex-bearing cells. Grafting experiments have not been possible with this family of plants because of the presence of laticifers. However, tracer-feeding experiments have laid the foundation for the biosynthetic pathway (Fig. 6.6), using whole plants and tissue cultures. Additional work of interest has been concerned with efforts to develop tissue culture systems for the production of morphine and codeine (Table 6.4). This work has provided evidence for the essential role of laticifers in the accumulation of these alkaloids, and also points to their possible involvement in morphine alkaloid biosynthesis.

Using sucrose density gradient centrifugation it has been possible to demonstrate that in exuded *P. somniferum* latex, the alkaloids are mainly located in dense vesicles. It has also been noted that the morphine:thebaine ratio in these organelles increases as capsule development progresses, an observation confirming the role of thebaine as an immediate biosynthetic precursor of codeine and morphine. In addition, the vesicles were shown to accumulate ^{14}C-methyl-morphine and ^{14}C-dopamine when these compounds were fed exogenously (Roberts *et al.*, 1988). Figure 6.7 shows an electron micrograph of latex vesicles isolated from *P. somniferum*. In subsequent investigations of *P. bracteatum*, thebaine which is the major alkaloid in this species was again found to be localized in dense vesicles deriving from laticifers. However, dopamine and sanguinarine were located in different subcellular fractions. Despite the great importance of these medicinal plants, it is quite typical that controversy still continues concerning the synthetic site for the morphine alkaloids; this is a true reflection of the present level of understanding of plant secondary metabolism and its regulation.

Fig. 6.7 (a) Latex vessel in *Papaver bracteatum* showing alkaloid-containing organelles and endoplasmic reticulum (near the periphery of the cell). Magnification × 13,500. (b) Alkaloid-containing organelles isolated from *Papaver somniferum* by spinning sucrose density gradients at 1000 g. (Courtesy of Dr M.F. Roberts and Dr E.M. Williamson, School of Pharmacy, University of London.)

(a)

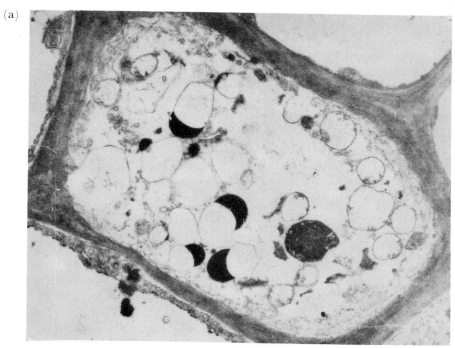

(b)

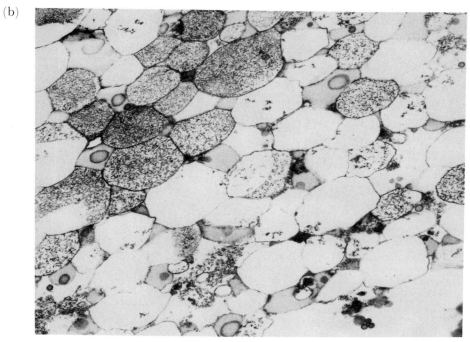

METABOLIC PATHWAYS AND THEIR REGULATION

The routes by which secondary metabolites are synthesized in plants are infinitely diverse and often intricate. In this brief section some pathways to products having relevance to the present discussion will be treated in some depth, and in particular the role of cell culture systems in determining these biosynthetic sequences will be highlighted. These examples have been chosen to illustrate how different secondary biosynthetic pathways may relate to primary metabolism.

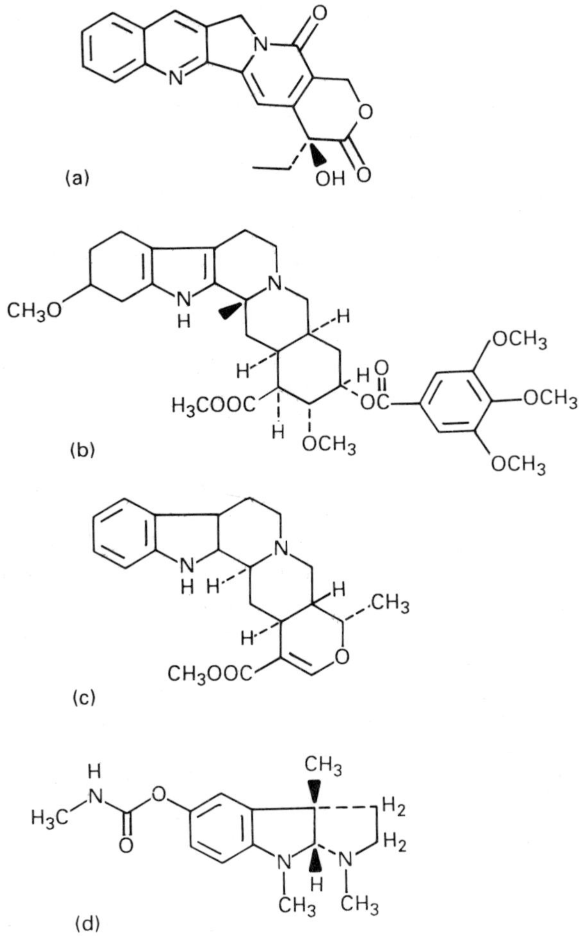

Fig. 6.8 Some indole alkaloids possessing important pharmaceutical activities. (a) Campothecin, an antineoplastic drug from *Camptotheca acuminata*; (b) reserpine, a tranquillizer from *Rauwolfia* species; (c) ajmalicine, an antihypertensive drug from *Catharanthus roseus*; (d) physostigmine, an anticholinesterase from *Physostigma venenosum*.

Indole Alkaloids

The biosynthesis of another class of alkaloids, the pyridine alkaloids of *Nicotiana* for example, was discussed on pp. 132–3. The primary precursors from which alkaloids are derived can differ widely, and the definitions of alkaloids in textbooks have been equally variable. However, most natural-product chemists would accept that any compound containing a nitrogen as part of a heterocyclic ring system could be described as alkaloid in nature.

Indole alkaloids, i.e. those derived from tryptophan and possessing an indole group, represent the largest group of all known alkaloids. More than 1200 tryptophan-derived alkaloids have been described, the major interest in this group dating from the early 1950s, when reserpine was isolated from *Rauwolfia serpentina*. The therapeutic properties of this compound, used as a tranquillizer, triggered the search for other indole compounds possessing valuable pharmacological activities. Figure 6.8 shows some of these.

The biosynthesis of one group of indole alkaloids, the so-called heteroyohimbine-type, was in part elucidated using cell-free systems prepared from cell suspension cultures; i.e., more or less crude homogenates prepared in such a way that enzyme activities are maintained. Several arguments have been put forward in this context for the use of suspension cultures rather than whole plants. First, cultures can be grown under controlled conditions and on a relatively large scale, allowing the production of biomass with relatively high homogeneity. In addition, in certain cases it may be possible to inhibit the action of enzyme inactivators such as phenolics, by growing suspension cultures under appropriate conditions. Phenolics are themselves secondary products, often accumulated by plants, but particularly when their tissues are traumatized.

One major limitation of plant cell suspension cultures is their inability to synthesize some secondary products to significant levels, a problem often attributed to their undifferentiated state (see p. 130). However, in the case of the heteroyohimbine-type indole alkaloids a well-defined growth medium was developed for cell suspensions of *Catharanthus roseus*, in which high yields of ajmalicine and serpentine were produced (Zenk *et al.*, 1977). Repeated selection for high alkaloid-producing cell strains resulted in suspension cultures which could produce higher alkaloid levels than those found in the source plants. These cultures were capable of accumulating serpentine, ajmalicine and small amounts of 19-epiajmalicine and tetrahydroalstonine, making them an excellent material for the study of enzymes involved in the biosynthesis of *C. roseus* alkaloids. Radioimmunoassays were developed in order to facilitate the detection of these alkaloids at picogram levels in crude tissue extracts (see p. 131 for further details).

This biosynthetic pathway now ranks as one of the best explored in plant secondary metabolism and, partly due to the ease with which *Catharanthus roseus* cell cultures can be manipulated, a fascinating picture of how alkaloid biosynthesis can be regulated is being built up. Studies in which precursors have been fed to cultures have confirmed theories that their flux through the two primary pathways leading to tryptamine and secologanin (Fig. 6.9) is of major importance to final alkaloid yield. Three enzymes in particular have been researched which appear to

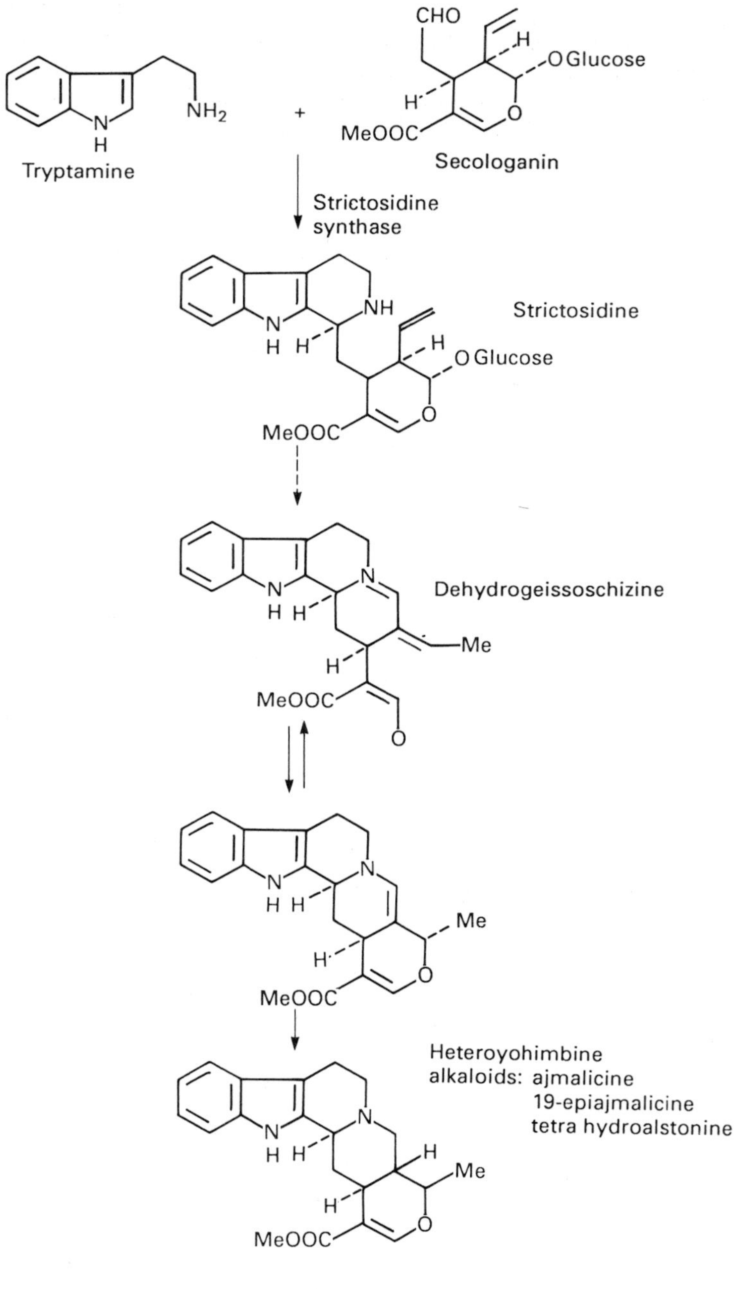

Fig. 6.9 Biosynthesis of heteroyohimbine alkaloids in *Catharanthus roseus*.

have important regulatory roles. These are geraniol hydroxylase, tryptophan decarboxylase, and strictosidine synthase. In some cell suspension lines, the activities of these enzymes have been enhanced on 'alkaloid-production' media, indicating that they may be induced under these conditions. The key role of strictosidine synthase is indicated in Fig. 6.9, and the conversions of tryptophan decarboxylase and geraniol hydroxylase are described in Fig. 6.10.

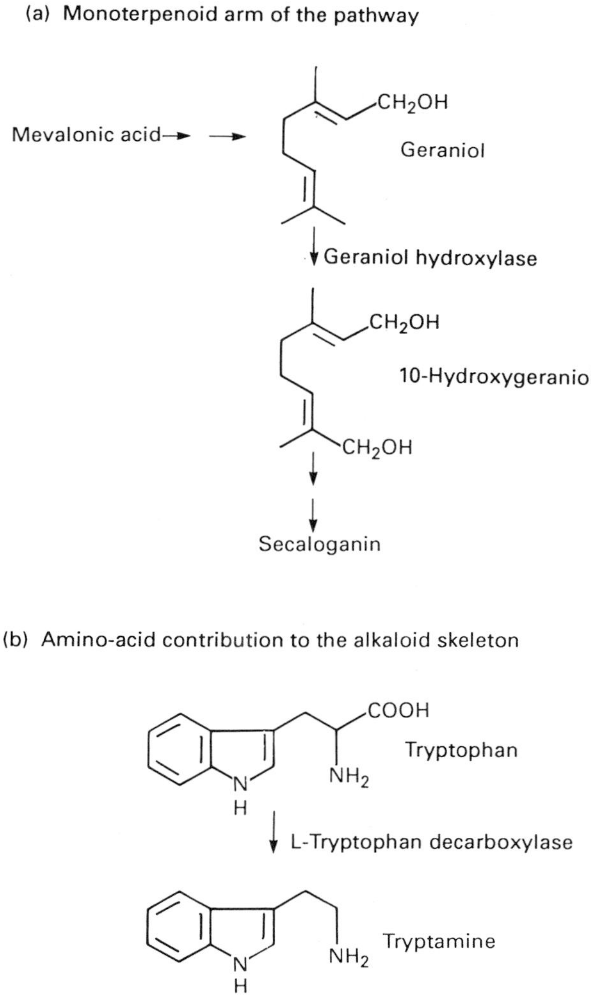

(a) Monoterpenoid arm of the pathway

(b) Amino-acid contribution to the alkaloid skeleton

Fig. 6.10 Key enzymic conversions in the biosynthesis of *Catharanthus* alkaloids. (a) Monoterpenoid arm of the pathway; (b) amino-acid contribution to the alkaloid skeleton.

These enzymes are likely to have an important role in the regulation of alkaloid pathways in many other species, strictosidine having been found to be a central precursor for the synthesis of monoterpenoid indole alkaloids in a number of plant families. Examples include strychnine from *Strychnos* species (family *Loganaciae*), camptothecin from *Camptotheca acuminata* (family *Nyssaceae*), and quinine alkaloids from *Cinchona* species (family *Rubiaceae*).

Flavonoids
Flavonoids constitute a class of phenolic compounds of very widespread occurrence in plants, some of the best known being a large proportion of flower pigments: anthocyanins (in angiosperms) and betalains (in *Centrospermae* and *Cactaceae*). The early steps in the biosynthesis of various types of flavonoid are closely related (Fig. 6.11), most of the pathway having been elucidated via the use

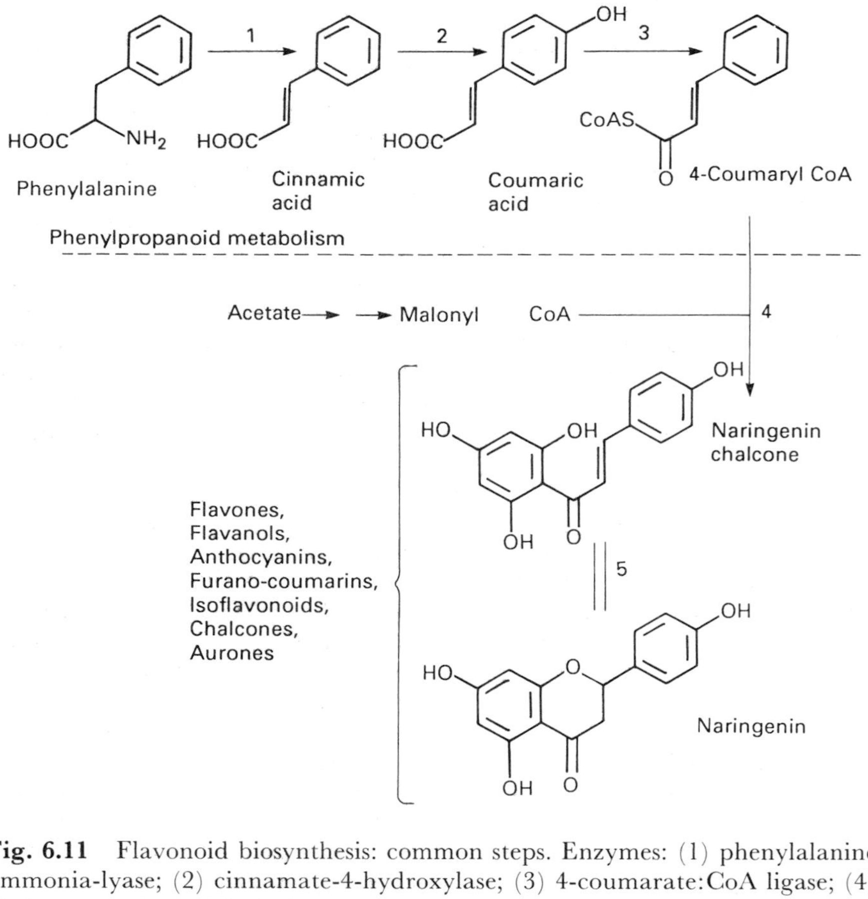

Fig. 6.11 Flavonoid biosynthesis: common steps. Enzymes: (1) phenylalanine ammonia-lyase; (2) cinnamate-4-hydroxylase; (3) 4-coumarate:CoA ligase; (4) chalcone synthase; (5) chalcone isomerase.

of radioisotope feeding. More recently, some key studies on the enzymatic regulation of these pathways have been undertaken, and have made important contributions to our understanding of how some secondary biosynthetic pathways may be regulated in response to external stimuli. These findings have in consequence been of considerable use to those workers trying to enhance secondary product synthesis in cell cultures for commercial purposes.

It has been known for several years that the expression of PAL activity is strongly correlated with flavonoid or furanocoumarin production in whole plant tissues and in plant cell cultures after treatment of these with a range of stimuli including UV light and fungal extracts. It has been suggested that flavonoids may in fact have a protective function for the plant in sunlight as these compounds absorb UV light very strongly. The analogous response to fungal extracts may be connected with the fungitoxic properties of furanocoumarins, which are also derived from the chemical intermediate chalcone.

Cell suspension cultures of parsley (*Petroselinum hortense*) were used to examine the effects of these stress conditions in more detail. Hahlbrock and co-workers (Chappell and Hahlbrock, 1984) found that the three enzymes of general phenylpropanoid metabolism ((1), (2) and (3) on Fig. 6.11) were co-induced in cultured parsley cells together with around 16 enzymes of the flavonoid pathway upon a treatment with UV light. When these cells were treated with an 'elicitor' derived from a fungal or microbial pathogen, the three common enzymes were co-induced with the furanocoumarin pathway. It was also shown that the enzymes were induced *de novo*, and that transient increases in the transcription rates of the genes coding for these enzymes could be detected.

Another group of workers studying the effects of an elicitor (a polysaccharide from fungal cell walls) on isoflavonoid phytoalexin formation in *Phaseolus vulgaris* have made similar observations (Cramer *et al.*, 1985). In this case the isoflavonoids being produced are phaseollin and kievitone, by similar co-ordinately induced parallel pathways. Accompanying changes include the deposition of wall-bound phenolic compounds and the production of hydroxyproline-rich glycoproteins, which also accumulate in the cell wall and may act to aggravate microbial pathogens or to strengthen the wall. However, this work has now progressed to the stage at which nearly all the enzymes on the isoflavonoid pathway have been characterized, and several of the genes have been cloned. The chalcone synthase gene in particular has been the subject of intensive research, and clones of this gene from a number of different species have been exploited as tools to examine new molecular approaches to the understanding and manipulation of plant biosynthetic pathways. In the particular system described above, Dixon and co-workers have extended their work to analyse the 'promoter' region of the chalcone synthase gene. They have identified regions which control transcriptional activation as well as the 'silencing' of the gene, and have demonstrated a nuclear repressor protein which binds to the silencer region. The ultimate outcome of this type of approach may be the development of promoters which as a result of genetic engineering would allow the co-ordinate induction of enzymes catalysing totally different biosynthetic pathways in cultured plant cells.

Inactivation of the chalcone synthase gene in the pigmented flower tissues of

Antirrhinum majus provided evidence for the activity of 'transposable elements' in this species. The apparently spontaneous mobility of these short stretches of DNA provides one possible mechanism of somaclonal variation in *Antirrhinum* and *Petunia*, and the phenomenon is likely to extend to other plant species.

Recently, a potentially very exciting approach to the manipulation of plant secondary biosynthetic pathways has been demonstrated using the chalcone synthase gene as a target. Van der Krol and co-workers (1988) used a clone of this gene in *Petunia* to produce an 'anti-sense' gene, i.e. an inverted copy of the gene spliced to a promoter which was in the normal orientation. This was then inserted into the plant by genetic transformation. Flowers of the regenerated plants possessed a range of colour phenotypes ranging from deep red to white, and genetic analysis of these revealed that transcription of the 'anti-sense' genes had at least partially inhibited the expression of the chalcone synthase gene. By extrapolation from bacterial systems, in which this strategy has been more widely explored, it is believed that the inhibition of gene expression in this case resulted from the synthesis of an 'anti-sense mRNA' from the inserted gene, which then formed a duplex with the correctly orientated mRNA produced from the natural chalcone synthase gene. Translation was thus inhibited and no, or only defective, protein products were formed.

The great significance of this approach is that it could in theory be used to prevent the synthesis of any secondary product, as long as the enzymes controlling the route were known and the target gene cloned. Possible applications are the prevention of turnover of 'target' secondary products in tissue cultures, and the inhibition of toxin synthesis in forage crops (e.g. cyanogenic glycosides, or alkaloids). Of course, it is not known how higher levels of product accumulating prior to the 'anti-sense block' might affect the overall biosynthetic rate; product feedback repression may become a problem. However, the importance of this approach cannot be disputed and further examples of its application should not be long in coming.

It should be realized that the production of these isoflavonoids in response to elicitor application is likely to be a special case; co-ordinate enzyme induction may be a feature peculiar to the production of defence metabolites in the face of predator or pathogen attacks.

Secondary Product Synthesis by Plant Tissue Cultures

PRODUCTS OBTAINABLE FROM TISSUE CULTURES

The investigation of secondary metabolite biosynthesis by plant cell and tissue cultures has gained momentum worldwide in recent years for two main reasons. One is that plant tissue culture techniques can now be applied to most species; although it has been estimated recently that only around 2000 different plant species have ever been grown in cultures, some kind of positive response can be expected for all species given a degree of patience and time to test variables such as culture media. The second reason is that the potential of some plant culture systems for the production of medically important compounds has been demon-

Table 6.5 Examples of secondary products reported from plant cell and tissue culture

Chemical class	Product	Species
Alkaloids	Atropine	*Atropa belladonna*
	Berberine	*Coptis japonica*
	Caffeine	*Camellia sinesis*
	Camptothecin	*Camptotheca acuminata*
	Harringtonine	*Cephalotaxus harringtonia*
	Nicotine	*Nicotiana tabacum*
	Papaverine	*Papaver somniferum*
	Sanguinarine	*Papaver somniferum*
	Vindoline	*Catharanthus roseus*
Steroids	Cholesterol	*Dioscorea deltoidea*
	Diosgenin	*Dioscorea deltoidea*
	Solasodine	*Solanum nigrum*
	Stigmasterol	*Brassica napus*
	Sitosterol	*Artemisia absinthium*
	Panaxadiol	*Panax ginseng*
	Tigogenin	*Trigonella occulta*
Phenylpropanoids	Anthocyanins	*Daucus carota*
	Shikonin	*Lithospermum erythrorhizon*
	Ubiquinone-10	*Nicotiana* spp.
	Anthraquinones	*Gallium mollugo*
	Capsaicin	*Capsicum annuum*

strated, and it can be said with certainty that these systems can provide a commercially realistic alternative to whole plants for the production of some drugs. A fraction of the list of secondary metabolites produced by plant cell cultures are given in Table 6.5. The table is deliberately brief, and it should be emphasized that many of the compounds mentioned can be produced by other species in culture, likewise most of the species listed are capable of the production of a wide range of secondary metabolites. Lastly, many other examples could be mentioned which fall into alternative chemical classes. However, tables of 'successes' such as this one should be regarded with caution. To begin with, no mention is made of those species examined which do *not* yield the compound in question (found in the whole plant) when cultured as cell suspensions. Second, the yields obtainable from the cultures cited have not been given; although yields comparable with or greater than those found in the whole plant can be and have been achieved (Table 6.6), it is usually only with considerable effort. Because the cell suspension culture has generally been considered to be the most adaptable form of plant tissue culture from the point of view of scale-up, until recently most emphasis has been placed upon attempts to induce specific secondary metabolisms in undifferentiated culture systems. Differentiated cultures are now attracting

Table 6.6 Yields of secondary products from cell suspension cultures and whole plants

Chemical product	Plant species	Culture type	Yield (% dry weight) culture	plant
Anthraquinones	*Cassia tora*	Callus	6	0.6
Glutathione	*Nicotiana tabacum*	Suspension	1.0	0.1
Ajmalicine	*Catharanthus roseus*	Suspension	1.0	0.3
Serpentine	*Catharanthus roseus*	Suspension	0.8	0.5
Rosmarinic acid	*Coleus blumei*	Suspension	15	3
Ginsenosides	*Panax ginseng*	Callus	27	4.5
Shikonin	*Lithospermum erythrorhizon*	Callus	12	1.5
Diosgenin	*Dioscorea deltoidea*	Suspension	2	2

attention, and some of these will be discussed below with reference to specific examples. However, the majority of the remainder of this chapter will concentrate upon the achievements in product synthesis by cell suspension cultures, and the methods used to induce and enhance yields will be considered in some depth.

Shoot Cultures
Shoot cultures often produce higher product levels and more conventional product profiles than their corresponding callus or suspension cultures. For example, in *Catharanthus roseus* shoot cultures, the dimeric alkaloids anhydro-vinblastine and leurosine have been detected, although at lower levels than those found in the leaves of the plant, while cell suspension cultures contained no trace of these complex structures. Similarly, shoot cultures of *Cinchona ledgeriana* contain much higher levels of quinine and related alkaloids than the cell suspension cultures, and *Digitalis* shoot cultures can accumulate more cardiac glycosides than undifferentiated cultures of this species.

One group of compounds not yet discussed in any depth is the lower terpenoids: i.e. the monoterpenoids and sesquiterpenoids which constitute a large part of the aroma and flavour chemicals produced in nature. These compounds and others related, which are often classified as 'essential oils', are frequently not synthesized in undifferentiated cultures, and this is generally considered to be a reflection of the lack of specialized compartments in these tissue cultures. Though it has been shown that many of these oils can occur in the form of glycosides, the aglycone (free) form is cytotoxic and must be accumulated in a compartment external to the cytoplasm. Table 6.7 provides a few examples where oil accumulation has been

Table 6.7 Essential oil production in plant tissue cultures

Species	*Compound(s)*	*Culture morphology*
Mentha piperita	Geraniol, linolool (early precursors)	Callus
Mentha piperita	Menthone, menthol (terminal products)	Callus with newly formed secretory glands
Origanum vulgare	Monoterpenoids	Shoots regenerated from callus
Lavandula angustifolia	Monoterpenoids	Shoots regenerated from callus
Eucalyptus citriodora	Monoterpenoids including citronellol	Root initiation required
Pimpinella anisum	Anethole	Cell suspensions in the presence of a lipophilic phase
Matricaria chamomilla	Sesquiterpenes including α-bisabolol	Callus containing idioblasts (oil cells) or cell suspensions in the presence of a lipophilic phase
Pelargonium spp.	Monoterpenes including menthone, limonene, phellandrene	Callus containing oil glands; shoot cultures

correlated with the appearance of a distinct culture morphology. Some chemical formulae are given in Fig. 6.12. In those cases listed, none or only very low levels of the oils found in the whole plant could be detected in undifferentiated cell cultures growing under normal conditions. However, the incorporation into the medium of a 'lipophilic phase' has on occasions been found to enhance accumulation. These phases may be liquid, as for example Miglyol, a neutral triglyceride, or solid, as in the case of ion-exchange resins such as XAD-40. The highest levels of essential oil have been observed when the culture differentiates the accumulating structures normally found in the whole plant. These may be specialized 'oil cells', or, more often, complex multicellular glands or glandular hairs may be necessary (Fig. 6.13). These appear to be required in the case of *Pelargonium*, for example, where Brown and co-workers have demonstrated that the formation of glands is closely correlated with essential oil formation (Brown and Charlwood, 1986). This species is particularly amenable to regeneration from callus, and these workers have been able to develop shoot proliferation cultures growing submerged in a simple bioreactor. The most important feature of the cultures is the numerous glandular hairs which they bear on the leaf surfaces. While the technology for growing shoot cultures is relatively undeveloped, the possibility of manipulating similar cultures of other species to produce products associated specifically with shoots or leaves is an exciting one, and deserves further attention.

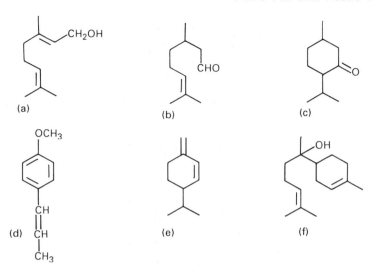

Fig. 6.12 Some plant essential oils (see Table 6.7). (a) Geraniol, (b) citronellol, (c) menthone, (d) anethole, (e) β-phellandrene, (f) α-bisabolol.

Fig. 6.13 Scanning electron micrograph of multicellular glands on a hop inflorescence (bracteole). These glands accumulate the flavour compounds of hop (mostly essential oils) important in brewing.

Fig 6.14 Transformed 'hairy roots' of *Datura stramonium* growing in a fermenter. (With thanks to Dr M. Rhodes, AFRC Institute of Food Research, Norwich.)

Root Cultures

In those cases in which roots constitute a main site of synthesis in the plant, root cultures should be considered as an alternative system to cell suspension cultures. Until recently, however, only a very limited number of plant species had been shown to yield root cultures which could be grown indefinitely and with an acceptable growth rate. The recognition of the useful effects of a long-known plant pathogen has altered our concept of root cultures. *Agrobacterium rhizogenes* is a soil bacterium capable of genetically transforming plant cells. The introduction of a small segment of so-called Ri-DNA into the plant genome during infection causes the expression of the 'hairy root' phenotype. The mode of transformation, and the use of this phenomenon in plant improvement is discussed further in Chapter 5 (p. 120). In the present context, the importance of hairy roots lies in a number of features. First, they are often capable of rapid and indefinite growth, with prolific lateral root formation, on medium containing no phytohormone (Fig. 6.14). Second, they have been shown to stably produce levels of products comparable to those normally found in the roots of whole plants. Species of many plant families have been successfully transformed using different biotypes of *A.*

Table 6.8   Hairy root cultures producing secondary products

Species	Family	Secondary products
Duboisia myoporoides	*Solanaceae*	Scopolamine
Atropa belladonna	*Solanaceae*	Atropine
Nicotiana rustica	*Solanaceae*	Nicotine, anatabine
Catharanthus roseus	*Apocynaceae*	Ajmalicine, serpentine, catharanthine
Tagetes patula	*Compositae*	Thiophenes
Lithospermum erythrorhizon	*Boraginaceae*	Shikonin
Beta vulgaris	*Chenopodiaceae*	Betacyanin

rhizogenes (Table 6.8), making these cultures yet another interesting alternative to cell suspension cultures. Several laboratories are now developing methods for growing these cultures in large volumes, and their successful cultivation in fermenters of up to 30 L has been reported. The major obstacle to the more widespread use of this approach seems to be the host-range of the wild-type *A. rhizogenes*. It is mainly non-woody dicotyledonous plants which can produce rapidly growing hairy roots following infection, and many of the more important pharmaceutical plants do not fall into this category. However, further research may resolve this problem.

Much of the research on the properties of hairy-root cultures for secondary product formation has been conducted using cultures of *Solanaceae* species, mainly *Nicotiana* and *Datura*. These plants accumulate pyridine and tropane alkaloids respectively, the biosynthetic pathways for these compounds having a common intermediate, *N*-methyl pyrroline (Fig. 6.15). The possibility that two enzymes in particular, putrescine methyl transferase and ornithine decarboxylase, may regulate the flux through these pathways, has led Rhodes and co-workers (1989) to tackle the cloning of the genes for these two enzymes with the objective of splicing the genes to a high-expression promoter (e.g. CaMV 35s promoter from cauliflower mosaic virus) and reinserting these back into cell cultures. The overall aim is the enhancement of biosynthetic activity and elevation of product yield. This approach provides a further example of how secondary metabolic pathways can be manipulated non-empirically, in both whole plant and cell and tissue cultures. So far, a yeast ornithine decarboxylase gene (ODC) has been inserted into tobacco hairy roots in an attempt to enhance the nicotine level. The next steps, characterization of these transgenic cultures, will provide some interesting information on the likelihood of this strategy succeeding.

Embryo Culture

Somatic embryo culture is an obvious target system if products are known to be embryo-produced and/or accumulated. The immature cotyledons of many plant species are the location of important compounds acting as anti-feedants or storage reserves for the germinating seed.

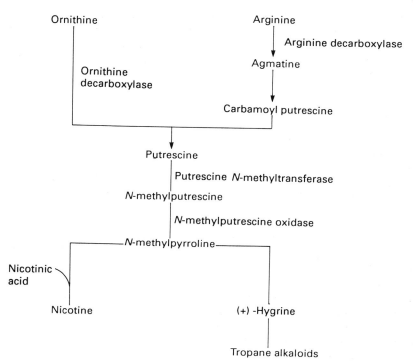

Fig. 6.15   Tropane and nicotine alkaloid pathways.

Somatic embryo cultures may be produced in a number of different ways. Direct somatic embryogenesis involves the production of asexual embryos directly from the source plant tissue, which may be zygotic embryo, cotyledon, or even leaf or stem tissue. Indirect somatic embryogenesis refers to the differentiation of embryos from undifferentiated callus or suspension cultures. From the aspect of investigating fundamental biosynthetic processes and developmental pathways, somatic embryos derived via any of these methods are potentially interesting experimental systems, although the synchronization of embryo development may present problems. However, from another angle, that of deriving a biotechnological process based upon somatic embryo culture, the only readily exploitable system is one which depends upon the differentiation of cells in a suspension culture. This type of differentiation usually requires the imposition of an environmental trigger such as alteration in hormonal status or carbohydrate levels. For instance, cell suspensions of carrot (*Daucus carota*), can be induced to produce somatic embryos by the removal of 2,4-D from the medium. However, only 'embryogenic' cell lines will respond well to this trigger. In general, the capacity for embryogenesis of any undifferentiated culture appears to be related to the age of the culture, and sometimes to the type of explant from which the culture was originally established.

Somatic embryo cultures are an obvious choice for the study of the accumul-ation of storage reserves such as lipids. In one study on *Papaver somniferum* and *P. orientale* cultures, efficient induction of somatic embryogenesis was effected by the removal of 2,4-D from the basal medium. Storage lipid (triacyl glycerol) accumulation preceded visible morphological changes, and was followed by an increase in sanguinarine concentration, particularly in *P. somniferum* cultures. Morphinane alkaloids appeared in the cultures about 45 days after induction of embryogenesis. Targets for strategies of this nature could be unusual seed storage lipids such as γ-linolenic acid. This unsaturated fatty acid has attracted interest because of numerous indications that it may have therapeutic effects in a range of physiological disorders, including premenstrual tension, migraine, multiple sclerosis, and allergic reactions. γ-Linolenic acid has most frequently been obtained from the oily seed of the evening primrose (*Oenothera biennis*) but has been detected at high levels in other plants, many belonging to the family *Boraginaceae*.

An additional target may be the production of cocao butter from embryos of *Theobroma cacao*. Jules Janick and co-workers at Purdue University have inves-tigated somatic embryogenesis in this species extensively, evaluating the culture conditions necessary for the continued development of the embryos, and changes in lipid, alkaloid, and pigment profiles during maturation (Janik *et al.*, 1982). One of the most critical factors affecting embryo maturation was the medium sucrose concentration, a stepwise increase of from 3 to 27% being required over 74 days of *in vitro* culture. During this time, lipid, alkaloid (theobromine and caffeine) and anthocyanin biosynthesis occurred, by which these workers inferred that the regulation of these metabolic routes was in part a common process. The lipid profiles obtained in these embryos resembled those found in commercial cocoa butter as used in the manufacture of chocolate.

SECONDARY PRODUCT ACCUMULATION BY PLANT CELL SUSPENSION CULTURES

There are many excellent reviews covering this field, for example Robinson (1981), Pais *et al.* (1988), Robins and Rhodes (1988), Stafford *et al.* (1986). Our main reason for concentrating on the subject of product accumulation by cell suspension cultures in particular is that these systems have attracted enormous interest largely because of their potential for scale-up. Some of the bioreactor design considerations for plant suspension cultures are discussed in Chapter 9. While it should be recognized that suspensions frequently do not achieve the secondary product yields required of them, and more often than not produce much less than the whole plant on a percentage dry weight basis, the fact remains that as yet, liquid suspensions are the easiest plant culture systems to initiate, maintain, and scale-up. A second very important point is that even given low product yields, the levels and activities of secondary biosynthetic enzymes in cultures may often be high relative to the whole plant, making cell cultures an excellent source of material for enzyme purification and investigations on the molecular regulation of biosynthetic pathways. In a few cases, the capacity of plant cell cultures to synthesize and accumulate secondary products has been remarkable (Table 6.9). Some strategies employed to enhance product yields in cultures will now be

Table 6.9 Examples of high secondary product yields achieved in plant cell cultures

Species	Compound	Maximum yield in culture (% dry weight)
Coleus blumei	Rosmarinic acid	23% (5.6 g product/L culture)
Lithospermum erythrorhizon	Shikonins	23% (6.4 g product/g inoculum) cf. plant 1–2%
Morinda citrifolia	Anthraquinones	10% (2.5 g product/L culture) cf. plant 1%
Catharanthus roseus	Serpentine	2%
Coptis japonica	Berberine alkaloids	15% (1.7 g product/L culture)
Panax ginseng	Gingsenoides	2% (150 mg product/L culture)
Nicotiana tabacum	Nicotine	2.1%

discussed, with particular reference to alkaloid production in cultures of *Catharanthus roseus*.

Yields Obtainable from Cell Suspension Cultures

Depending on the product type, the plant species, and the cell line, plant cell suspension cultures vary enormously in their capacity to produce and accumulate secondary products. With the advent of readily available genetic engineering technology, some exciting possibilities are arising for the directed manipulation of secondary biosynthesis in plant cell cultures. However, the vast majority of effort has been and still is being devoted to an empirical approach to the enhancement of product yield. Methods employed to achieve this objective are numerous, and some of these are listed in Table 6.10. Of these approaches listed, the adjustment of phytohormone type and quantity often has one of the most dramatic effects, and is certainly a variable to be assigned a high priority for investigation. When attempting to optimize culture conditions for the production of a given target compound, it is important to select a range of variables to manipulate: vital clues to which types of medium component are worth manipulating first can often be obtained by a critical perusal of available literature, bearing in mind the chemical nature of the target compound and the plant family. Specific applications of these strategies will be mentioned where appropriate in the examples to follow.

The screening and selection of high-yielding cell lines is an additional approach to the optimization of a commercial process, which can be adopted in parallel with environmental manipulation. The ultimate limitations to these approaches are the number of hands available to perform the work, and the space available to maintain the cultures; as a result screening programmes are often adopted subsequent to the development of suitable 'product induction' media. Strategies for the improvement of plants via cell culture have already been dealt with to some extent (Chapter 5), and methods for the analysis of secondary products at a micro-level in plant tissues have been briefly discussed earlier in the present chapter (radio- and enzyme-immunoassays; p. 131). While immunoassays are in

Table 6.10 Empirical methods for the manipulation of secondary product yield in plant cell suspensions

Carbohydrate source	(a) Type (sucrose often the best)
	(b) Concentration (an increase is often beneficial)
Nitrogen source	(a) Type (ratio of ammonia:nitrate, inorganic *vs.* organic)
	(b) Concentration (lowering the level enhances alkaloids in some cases)
Phosphate level	High phosphate is often inhibiting to alkaloid accumulation
Phytohormones	Adjustments to auxin and cytokinin level and type can have dramatic effects upon product yield
Light regime	Some cultures only grow in the dark; light is often a critical factor in secondary-product accumulation
Temperature	Lowering the incubation temperature often enhances product yield but depresses growth rate
Osmotic stress	Imposed by high sugar, high salt; often enhances yield
Precursors	Variable effects; often positive, but cell-line dependent
Elicitation	Treatment of cultures with autoclaved filtrates of fungal cultures often causes an increase in product level

theory the ideal method for the sensitive and specific quantitation of chemicals in small tissue samples, problems may arise to limit their application. For instance, most plant secondary products have molecular weights so small that they must be linked to a protein such as bovine serum albumin in order to generate a good immune response in the antibody-producing animal. Chemical linkage may in itself be difficult; and the antiserum produced may not always be highly specific. Cross-reactivity of antisera to other components in the crude plant tissue extract may also occur. However, such methods have been used with some success, and one of these cases is discussed below (pp. 156–7). Screening for high yield is obviously simplest in the case of pigmented products, when the human eye is the analytical tool.

Selection methods similar to those discussed in Chapter 5 have also been employed to try to improve the production characteristics of plant cell cultures, and one such example, that of alkaloid-yield enhancement in *C. roseus* will be mentioned below (pp. 156–7).

The range of product yields obtainable from plant cell suspensions is illustrated by the data in Table 6.10.

Alkaloids from C. roseus
The plant family to which *Catharanthus roseus* belongs (the Apocynaceae), is well-known for the wide range of indole alkaloids which the member species produce

(Robinson, 1981). *C. roseus* has attracted particular interest because the leaves of the whole plant accumulate the dimeric alkaloids vinblastine and vincristine, but only to low levels (*c.* 0.0005% of the dry weight). Because of their relative scarcity and their application in tumour therapy, these alkaloids have a market value as high as $1–3 million/kg, making them excellent targets for alternative production technologies such as plant tissue culture. Despite this fact, and in addition the relative ease with which cell and tissue culture systems of this species can be developed, the total biosynthesis of the dimeric alkaloids *in vitro* has never yet been achieved. However, a combination of research effort worldwide for over two decades has led to some exciting advances. A consideration of some of the approaches used to achieve high alkaloid production in cell suspension cultures of *C. roseus* will serve to illustrate strategies and principles which can be applied to any tissue culture system.

While the valuable dimeric alkaloids have proven elusive in *C. roseus* cell cultures, many other 'monomeric' indole alkaloids can be accumulated to high levels in certain cell lines and, with selection, cell culture yields exceeding those of the whole plant have been achieved. The pioneer work of Zenk and his colleagues in the late 1970s introduced *C. roseus* as a model species for secondary product-accumulating tissue cultures. These workers concentrated their efforts on the enhancement of serpentine and ajmalicine levels in their cultures. Both these alkaloids have quite high market values; ajmalicine, at around $2000/kg is used in the treatment of circulatory diseases, and is normally obtained from the dried roots of *Rauwolfia serpentina*. However, serpentine, which upon reduction with boro-hydride yields ajmalicine with high efficiency, is found in relatively high levels in plants of *C. roseus*. Cultures of these species had until this time only been found to produce low levels of a range of different alkaloids, these being assigned tentative identification on the basis of TLC, and by spraying with alkaloid-specific colour reagents such as Dragendorff's.

The most fundamental approach that Zenk and co-workers used to improve alkaloid production in their cultures was to examine the effect of growing their cell suspensions on a range of different media, varying the basal salts and the hormone composition, and also testing out a number of indole alkaloid precursors (Zenk *et al.*, 1977). The outcome of many experiments was the development of an 'alkaloid-production' medium for *C. roseus*, which contained the salts, vitamins, and iron supplements extracted from three different published media formulations, with 5% (w/v) sucrose, IAA, 6BA (6-benzyladenine), and L-tryptophan being added. Using this medium, serpentine yields of 162 mg/L and ajmalicine yields of 264 mg/L were obtained. However, high alkaloid yields were only achieved during the first transfer into production medium. This type of experimental strategy underlies the principle of 'two-stage' production processes, in which the cell biomass is bulked up under optimum growth conditions, followed by transfer to new culture conditions under which the biosynthesis of the target products is triggered. Alternative 'one-stage' systems have also been developed for *C. roseus* in which media supporting both growth and high alkaloid production are used.

Cell line differences should be mentioned at this point. While in general, for a given species in culture, particular culture conditions might be expected to produce a given effect with respect to growth and secondary product accumulation, this is not always the case. For instance, different selected cell lines of *C. roseus* have been found to produce widely different types and quantities of indole alkaloids when grown under exactly the same conditions. Equally striking are the effects of feeding indole alkaloid precursors such as tryptophan and secologanin to cell suspension cultures. These substrates can have positive or negative effects upon secondary product synthesis, presumably reflecting differences in intermediate pool sizes and in the feedback regulation of key enzymes in the biosynthetic pathway.

Many workers have attempted to utilize the underlying variation inherent in callus and suspension cultures in order to select high-yielding cell lines. This overall approach provides an additional route to the improvement of productivity, and though time-consuming and tedious in nature, it can often reap considerable reward. One method of screening for high-producing cell lines was again first reported by Zenk's group. They developed radioimmunoassays for ajmalicine and serpentine. The method was specific for these two alkaloids, and the sensitivity was such that the average alkaloid content of a single *C. roseus* root cell could be determined. Using as starting material a cell line which consistently produced higher than average yields of alkaloids, the cells were plated in agar production medium, incubated for up to two months, and the small colonies arising on the plates screened by radioimmunoassay for alkaloid content. Massive variation was found in the level of production; for both serpentine (0–1.4% dry weight) and ajmalicine (0–0.8% dry weight). The general strategy is shown in Fig. 6.16. Cell suspensions selected using this method were able to produce higher alkaloid levels than the whole plant.

An additional means of screening for serpentine accumulation is possible because of the fluorescent nature of the alkaloid. Under long-wave UV light, serpentine autofluoresces bright blue, and can be visualized within the cell vacuole using a UV light microscope. Relatively high-serpentine cell lines have been selected manually on the basis of high fluorescence, and the possibility of using automated apparatus such as FACS to sort cells has also been investigated.

An additional approach to the positive selection of high-alkaloid producing cell lines has been the application of a selection pressure in the form of a toxic amino-acid analogue. Most plant alkaloids are derived from one or more amino acids. In the case of the *C. roseus* indole alkaloids the amino-acid contribution to the alkaloid skeleton is derived from tryptophan, via tryptamine. 5-Methyl tryptophan is recognized by tryptophan-metabolizing enzymes as a substrate, and under normal conditions will kill off the treated cells. However, if a few cells in a population of several million have high internal tryptophan pools, these may compete with low levels of 5-methyl tryptophan to allow the survival of the cell population. Cell cultures derived from the application of such analogues are therefore enriched for the normal substrate, and have often been found to accumulate amino-acid levels orders of magnitude higher than normal. However, while the concept of increasing precursor level to enhance flux towards alkaloid biosynthesis may be attractive, in the case of *C. roseus* indole alkaloids this approach has rarely worked in practice. It may be that the monoterpenoid arm of the pathway is in fact the crucial one in

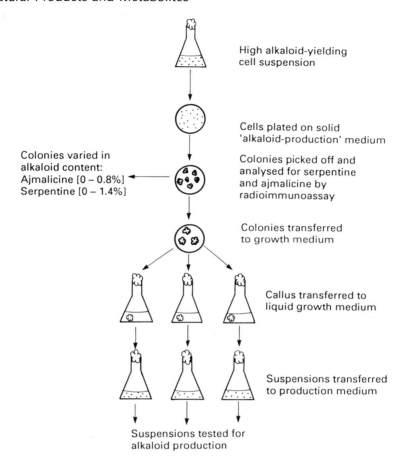

High alkaloid-yielding
cell suspension

Cells plated on solid
'alkaloid-production' medium

Colonies varied in
alkaloid content:
Ajmalicine [0 – 0.8%]
Serpentine [0 – 1.4%]

Colonies picked off and
analysed for serpentine
and ajmalicine by
radioimmunoassay

Colonies transferred
to growth medium

Callus transferred to
liquid growth medium

Suspensions transferred
to production medium

Suspensions tested for
alkaloid production

Fig. 6.16   Screening of *Catharanthus roseus* cell suspensions using radioimmunoassay.

determining the level of indole alkaloid accumulation in *C. roseus*, or that the limiting steps occur beyond the synthesis of tryptophan.

   C. roseus alkaloid-producing cell suspensions have been a most popular experimental system since the 1970s, with the result that the conditions favouring alkaloid accumulation have been extensively investigated. Table 6.10 lists some of the empirical methods which have been used to manipulate general secondary product yield in plant cell suspensions. Many of these apply to the specific case under discussion. Most important for the development of *C. roseus* alkaloid production media have been the use of high carbohydrate concentrations, decrease in phosphate content, and adjustment of phytohormones. To take each point in turn, some workers have used as an ajmalicine production medium an 8%

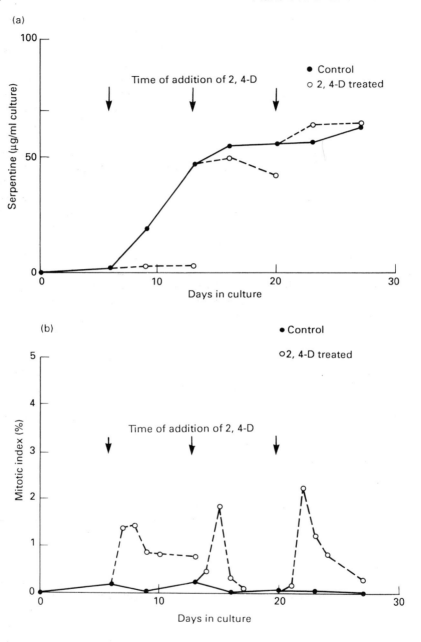

Fig. 6.17   Effect of addition of the auxin 2,4-D to a cell suspension culture of *Catharanthus roseus*. (a) Suppression of alkaloid accumulation; (b) increased mitotic activity.

sucrose solution; though biomass accumulation was low, alkaloid and phenolics accumulation was high. Equimolar concentrations of glucose, maltose and lactose have also been found to enhance alkaloid production in *C. roseus* cell suspensions. Phosphate is rapidly taken up by plant cells, stored in the vacuole and then gradually used to synthesize nucleic acids, phospholipids, and other organic molecules including ATP. While high initial phosphate levels have been found to have no effect upon tryptophan decarboxylase activity, they severely repress alkaloid accumulation, as well as the production of phenolics. With respect to the effects of phytohormones on alkaloid yield, the synthetic auxin 2,4-D., which is often included in callus induction and growth media, has a dramatic inhibitory effect upon indole alkaloid production. The mechanism of this action is unknown; 2,4-D has a number of effects when added to plant cell cultures, including rapid induction of mitotic acivity (Fig. 6.17), induction of the synthesis of new messenger RNA species, and in *C. roseus* 2,4-D has been found to inhibit the activity of tryptophan decarboxylase and strictosidine synthase. The transfer of cells from a 2,4-D-containing medium to a non-2,4-D medium brings about a dramatic alteration in the ability to accumulate alkaloids, usually after a lag phase of 3–6 days.

Zenk found that a tryptophan feed brought about enhanced alkaloid production in his *C. roseus* cell lines; other workers have detected the opposite effect, while the addition of secologanin, contributing the monoterpenoid portion of the indole alkaloid skeleton, is often observed to bring about increased alkaloid accumulation. Precursor-feeding studies such as these have in more recent years led to the study of key enzyme activities under alkaloid-production conditions, and very recently to attempts to clone the genes coding for these enzymes. Much effort has been directed towards the enzymes tryptophan decarboxylase and strictosidine synthase, with the result that both have now been characterized and the genes cloned. However, the regulatory role of these enzymes with respect to indole alkaloid production remains in some doubt, as their patterns of activity do not always reflect indole alkaloid accumulation profiles. In many studies, however, there has been no attempt to monitor alkaloid turnover, which is of course a major factor influencing the final level detected in cultures.

More consistent data point to a major regulatory role of the monoterpenoid arm of the indole alkaloid pathway. The enzyme geraniol-10-hydroxylase (see Fig. 6.10) is induced when cells are transferred to alkaloid production media, and its activity may therefore be a limiting factor in the biosynthesis of secologanin.

The main pharmacological interest in *C. roseus* as a plant species resides in its production of the dimeric alkaloids, vinblastine and vincristine. These compounds have never been detected in cell suspension cultures, but the final steps of the biosynthetic pathway have been studied in depth. A possible pathway is given in Fig. 6.18, largely based upon the results of numerous experiments using cell-free systems prepared from whole plants or tissue cultures of *C. roseus*. The transformation of tabersonine to vindoline is multi-step, and occurs in light-grown aerial tissues in the whole plant. A 22% efficient conversion of vindoline and catharanthine to 3′,4′-anhydrovinblastine using cell-free systems prepared from *C. roseus* tissue cultures was reported by Endo and his colleagues in 1986, followed by

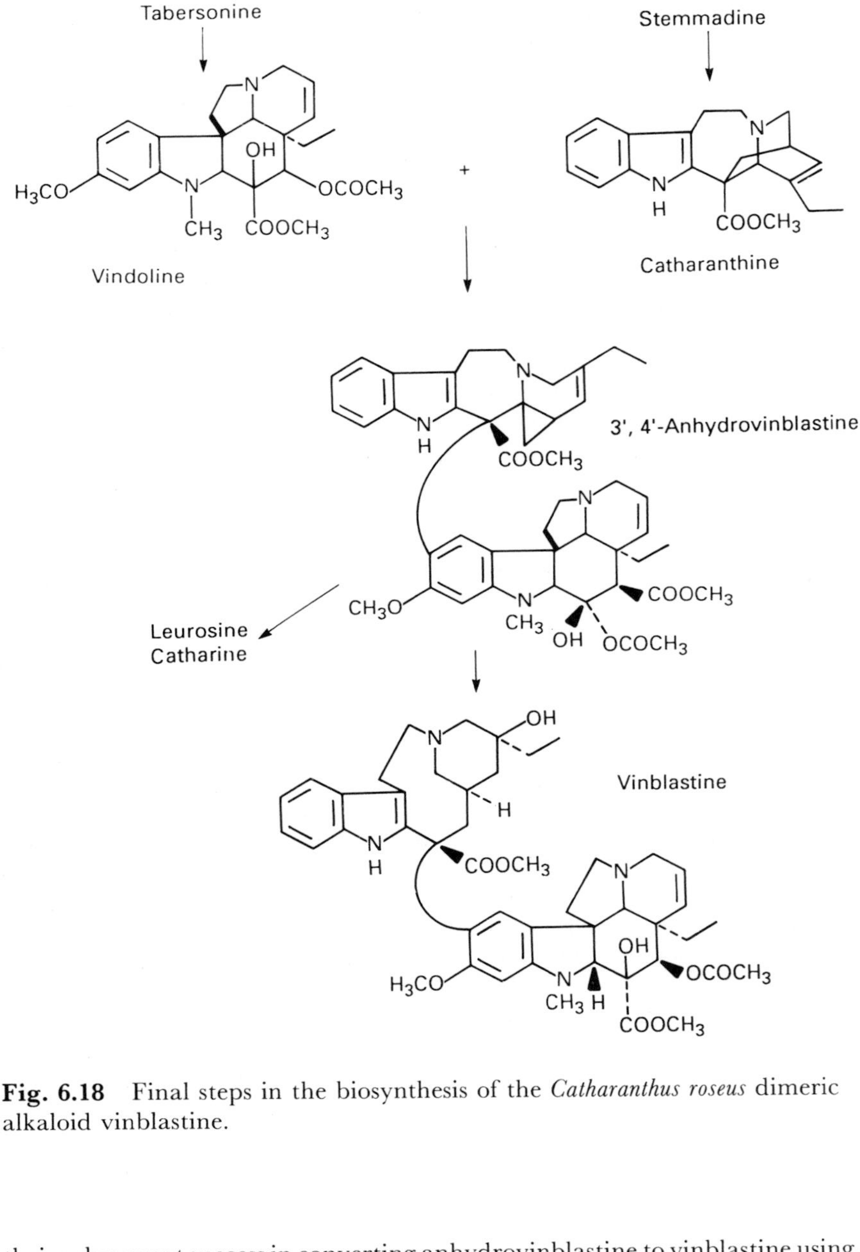

Fig. 6.18 Final steps in the biosynthesis of the *Catharanthus roseus* dimeric alkaloid vinblastine.

their subsequent success in converting anhydrovinblastine to vinblastine using cell suspension derived cell-free systems. In principle, the entire vinblastine biosynthetic pathway can now be achieved with the use of either whole *C. roseus* cell cultures or cell-free systems derived from these.

Table 6.11   Examples of novel compounds detected in plant cells

Compound	Plant species	Biological activity
Echinatin	*Glycyorhiza echinata*	—
24-Methylene cholesterol	*Holarrhena antidysenterica*	—
Lucidin	*Morinda citrifolia*	—
Pericine	*Picralima nitida*	CNS-active alkaloid
β-Peltatin A-Methylether	*Linum flavum*	Cytotoxic lignan (podophyllotoxin)
Rutacultin	*Ruta graveolens*	—

Novel Compounds

The wide-ranging potential of plant cell and tissue cultures for the production of known plant secondary products has been dealt with above. However, any plant cell culture may also provide a source of novel compounds, entirely unknown in the parent plant. A number of these are given in Table 6.11.

Given the diversity of secondary metabolic routes exhibited by even closely related plants, it is not surprising that undifferentiated tissue cultures derived from these should express different biochemical potentials. Some pathways may be completely repressed under culture conditions which favour the flux of primary precursors to alternative biosynthetic products. Thus, while plant cell cultures with more or less the same genetic potential as the parent plant may achieve new product profiles, genetically variant cell cultures that may have arisen during long-term maintenance would be expected to show even greater diversity.

With increasing interest in the pharmacological potential of plants, plant cell and tissue cultures should also come under much closer scrutiny. Sensitive biological assays are required in order to screen for the various activities of medicinal or agrochemical interest, for instance antiviral, antifungal, CNS-active or cytotoxic, but once these methods have been developed there is no good reason to restrict their use to the screening of whole plants. While secondary product yields in cultures maintained upon media designed for rapid growth might as a general rule be expected to be low, the chances of improving such levels when detected using a biological assay are good. This is an area of research yet to be investigated, however, and it requires the interest of pharmaceutical companies and other scientific organizations already active in screening for synthetic or natural drugs in order to develop further.

References and Further Reading

Brown, J.T. and Charlwood, B.V. (1986). The control of callus formation and differenti-ation in scented Pelargoniums. *Journal of Plant Physiology* **123**, 409–417.

Chappell, J. and Hahlbrock, K. (1984). Transcription of plant defence genes in response to UV light or fungal elicitor. *Nature* **311**, 76–81.

Cramer, C.L., Bell, J.N., Ryder, T.B. *et al.* (1985). Co-ordinated synthesis of phytoalexin biosynthetic enzymes in biologically-stressed cells of bean (*Phaseolus vulgaris* L.) *The EMBO Journal* **4**, 285–289.

Endo, T., Goodbody, A., Vukovic, J. *et al.* (1986). Enzymatic synthesis of 3′,4′-anhydrovinblastine by cell free extracts from cultured *Catharanthus roseus* cells. In *VI International Congress of Plant Tissue and Cell Culture* (D.A. Somers *et al.*, eds) IAPTC, Minneapolis, p. 143.

Harborne, J.B. (1973). *Phytochemical Methods, A Guide to Modern Techniques of Plant Analysis*. Chapman and Hall, London.

Janick, J., Wright, D.C., and Hasegawa, P.M. (1982). *In vitro* production of *Cacao* seed lipids. *Journal of the American Horticultural Society* **107**, 919–922.

Pais, M.S.S., Mavituna, F., and Novais, J.M. (eds) (1988). *Plant Cell Biotechnology*. Springer-Verlag, Berlin.

Rhodes, M.J.C., Robins, R.J., Aird, E.L.H. *et al.* (1989). Regulation of secondary metabolism in transformed root cultures. In *Proceedings of the Second International Symposium on Primary and Secondary Metabolism in Plant Cell Cultures, Saskatoon, 1988* (W.G.W. Kurz, ed.) Springer-Verlag, Berlin.

Roberts, M.F., McCarthy, D., Kutchan, T.M., and Coscia, C.J. (1988). Localization of enzymes and alkaloidal metabolites in *Papaver* latex. *Archives of Biochemistry and Biophysics* **222**, 599–609.

Robins, R.J. and Rhodes, M.J.C. (eds) (1988). *Manipulation Secondary Metabolism in Culture*. Cambridge University Press, Cambridge.

Robinson, T. (1981). *The Biochemistry of Alkaloids*, 2nd edn. Springer-Verlag, Berlin.

Staba, E.J. (ed.) (1980). *Plant Tissue Culture as a Source of Biochemicals*. CRC Press, Boca Raton.

Stafford, A., Morris, P., and Fowler, M.W. (1986). Plant cell biotechnology: a perspective. *Enzyme and Microbial Technology* **8**, 577–640.

Van der Krol, A.R., Lenting, P.E., Veenstra, J. *et al.* (1988). An anti-sense chalcone synthase gene in transgenic plants inhibits flower pigmentation. *Nature* **333**, 866–869.

Zenk, M.H., El-Shagi, H., Arens, H. *et al.* (1977). Formation of the indole alkaloids serpentine and ajmalicine in cell suspension cultures of *Catharanthus roseus*. In *Plant Tissue Culture and its Biotechnological Application* (W. Barz, E. Reinhard, and M.H. Zenk, eds), Springer-Verlag, Berlin, pp. 27–43.

Chapter 7

Biotransformation by Plant Cell Cultures

GAGIK STEPAN-SARKISSIAN

Introduction

Biotransformation is a process through which the functional groups of organic compounds are modified by living cells. Like many other processes currently used with plant cell suspension cultures, biotransformation was initially investigated with, and applied in, microbes.

Microbial systems provide a vast range of advantages for transformation processes. The number and diversity of microbial species with their associated wide range of potential for enzymatic catalysis, the relative ease with which they can be grown in large volumes, and their high rates of growth and metabolism are factors which make microbes very efficient agents for biological transformation. This potential of microbes was unwittingly harnessed by man almost at the dawn of history. In Mesopotamia for example the acetification of wine into vinegar was already known in the sixth millenium BC.

In the past 25–30 years significant advances have been made in the field of plant cell suspension cultures. There are nowadays hundreds of plant species in culture in different centres throughout the world. Suspension cultures of a few species like the Madagascan periwinkle (*Catharanthus roseus*) have been extensively studied in terms of their growth and primary and secondary metabolism. The potential of plant cell cultures for biotransformation processes was realized soon after their establishment, and they are used widely as tools for biosynthetic studies.

Because of factors such as slower rate of growth, difficulty of maintaining culture sterility, etc. plant cell cultures cannot compete with microbial systems for catalysis of the same biotransformation reaction. However, the plant kingdom is genetically very diverse and possesses a correspondingly rich repertory of enzymes.

From an industrial point of view, biotransformations performed by plant cell culture systems can be desirable when a given reaction is unique to plant cells and the product of the reaction has a high market value.

A major disadvantage of plant cell cultures compared with the whole plant in connection with secondary metabolism is that the biosynthetic potential of the species in question is very often not expressed in culture. Although the production of desirable secondary substances from cell cultures is considered a potential alternative to whole plant sources, at present total synthesis of complex biochemicals such as medicinal compounds from simple, primary precursors supplied in the medium is rarely achieved at a commercially significant level in suspension cultures. Harnessing the biotransformation potential of cell culture has therefore been a way of circumventing this drawback. In cases where the 'final' desirable product of a given pathway is not available from cells cultivated *in vitro*, due to the non-expression of one or more enzymes in the intermediate stages of the pathway, the supply of immediate or near precursors and their conversion into the 'final' product can be of advantage. Alternatively, the biotransformation of synthetic substances, analogues, or secondary metabolites from other plant species can result in novel compounds hitherto unknown in nature which may have new and useful biological properties.

Plant cells may be used in a number of ways for biotransformation purposes. The most basic procedure is to supply the cell suspension with the compound to be transformed and harvest the product from the culture (cells and medium) after an appropriate interval. Alternatively, plant cells possessing the potential of a given biotransformation reaction may be immobilized in an appropriate matrix. In this case the recovery of the product will be from the bathing medium. Finally, immobilized enzymes may also be used to perform biotransformation reactions. Under this option both whole plant tissue and plant cells cultivated *in vitro* may be used as sources of enzyme protein.

The presence of a biotransformation potential in plant cells is a necessary condition for a practical application. However, for a successful and viable process to be envisaged, other conditions must be met. These conditions can be summarized as follows:

- The substrate of a biotransformation reaction must be easily assimilated by the cell and reach the appropriate cellular compartment or organelle without significant degradation
- The substance must not be toxic to the cell culture
- The rate of product formation must significantly exceed the rate of its further metabolism.

This chapter discusses various processes involving biotransformation by plant cell cultures and plant enzymes and their biochemical potential. Emphasis is placed upon biotransformation reactions which result in compounds of greater complexity, usually secondary metabolites. Degradative-type biotransformation will be mentioned only when illustrating a point.

Biochemical Potential

Plant cell suspension cultures consist of undifferentiated cell aggregates and single cells growing in a nutrient liquid medium. The medium is a buffered solution of various organic and inorganic salts, a carbon source, and growth hormones (e.g. auxins and cytokinins). The selection of hormones for suspension cultures is of key importance, because these substances on their own or in combination with each other may cause differentiation or dedifferentiation of a culture and directly affect secondary metabolism by repressing and derepressing the production of certain key enzymes.

The original reasons for initiation of plant cell suspension cultures were mainly academic. These cultures were considered to have a number of advantages over whole plants for biosynthetic and physiological studies. Cell cultures have a minimum of organization and are undifferentiated systems in which precursor incorporation and fate can be monitored relatively easily. In addition, cell cultures can be grown on fully defined media under standardized conditions, and the aseptic environment in which cultures must necessarily grow ensure that biosynthetic reactions are catalysed solely by plant cells and not by contaminating microflora. Finally, cultured plant cells have short growth cycles and are not influenced by seasonal variations.

Soon after the techniques of initiating and growing plant cell cultures were established and their basic requirements for growth were understood, their potential for the production of secondary metabolites was also appreciated. Pigmentation and the presence of extracellular products and enzymes were indications that drew attention to this particular aspect in suspension as well as in static cultures of plant cells.

Instances where plant cell cultures have been reported to produce levels of secondary metabolites comparable with those of the parent plant are rare. The reason for this is generally ascribed to the low levels or complete absence of one or more enzymes in biosynthetic routes leading to secondary products. Plant cells are generally considered to be totipotent, i.e. they carry in them the complete set of genetic information present in the parent (although see Chapter 2). Two examples involving different approaches will be considered to illustrate this point.

When cell cultures lacking the ability to produce natural substances seen in the parent plant are used to regenerate the whole plant, the latter is found to have regained the ability to biosynthesize species-specific secondary metabolites. Alternatively, when cultures are initiated from parts of a plant not synthesizing a metabolite produced in another organ of the same plant, they sometimes produce and accumulate the metabolite. Despite this totipotency in undifferentiated plant cells, the genetic information in them is often not fully expressed. This characteristic is a major disadvantage if plant cell cultures are to be used in mass cultivation to produce desirable biochemicals. In certain cultures, the presence of metabolic defects and blocks can nevertheless offer certain advantages. The absence of an enzyme in a biosynthetic route results in the accumulation of the intermediates preceding that step, thus facilitating their isolation and characteriz-

Table 7.1 Some examples of reactions and chemical groups involved in biotransformations by plant cell suspension cultures (Kurz and Constabel, 1979)

1. Reduction	*2. Oxidation*
$C{=}C \rightarrow CH_2{-}CH_2$	$CH_3 \rightarrow CH_2OH$
$C{=}C{-}CO \rightarrow CH_2{-}CH_2{-}CO$	$CH_3 \rightarrow COOH$
$\rightarrow CH_2{-}CH_2{-}CHOH$	$CH_2OH \rightarrow CHO$
$CO \rightarrow CHOH$	$CHOH \rightarrow CO$
$CHO \rightarrow CH_2OH \rightarrow CH_3$	$CH_2{-}CH_2 \rightarrow CH{=}CH$
	$CH_2{-}CH_2{-}NH_2 \rightarrow CH_2COOH$
	$={S} \rightarrow S{=}O$
3. Hydroxylation	*4. Epoxidation*
$CH \rightarrow C{-}OH$	
$CH_2 \rightarrow CH{-}OH$	$CH{=}CH \rightarrow HC\overset{\displaystyle O}{\underset{}{\diagdown\diagup}}CH{-}$
$NH_2 \rightarrow NH{-}OH$	
5. Glycosylation	*6. Esterification*
$OH \rightarrow O\text{-Glucose}$	$OH \rightarrow O\text{-Ac}$
$OH \rightarrow O\text{-Rhamnose}$	$\rightarrow O{-}COCH_2NEt_2$
$OH \rightarrow O\text{-Apiose}$	$\rightarrow O\text{-Palmitate}$
$COOH \rightarrow COO\text{-Glucose}$	$\rightarrow O\text{-Malonate}$
$CH \rightarrow C\text{-Glucose}$	$\rightarrow O\text{-Succinate}$
$N \rightarrow \equiv N^+\!{-}\text{Arabinose}$	$COOH \rightarrow COO{-}\text{Malate}$
7. Methylation and demethylation	*8. Isomerization*
$OH \leftrightharpoons O{-}CH_3$	*trans* $\leftrightharpoons$ *cis*
$\equiv N \rightarrow \equiv N^+\!{-}CH_3$	$D \rightarrow L$
$\equiv N^+\!{-}CH_3 \rightarrow {=}NH$	$\beta\text{-OH} \rightarrow \alpha\text{-OH}$

ation. In certain cases, this aspect also provides opportunities for biotransformation reactions.

Plant enzymes can be grouped into two classes, depending on whether their specificities are directed towards particular substrates or functional groups. The first group of enzymes will naturally reject compounds which do not fulfil substrate requirements and are unnatural. It is the second group of enzymes, such as alcohol dehydrogenases, glycosidases, and peroxidases, which offer the most potential for bioconversion reactions and are able to transform unnatural substances.

For nearly two decades plant cell cultures have been used for the transformation of important classes of compounds such as cardiac glycosides, alkaloids, and steroids. The range of reactions and that of chemical groups involved is wide (Table 7.1) and the scope for biotransformations appears to be limited only by the diversity of natural and unnatural substrates available for such reactions (Kurz and Constabel, 1979).

The biochemical potential of plant cell cultures may be exploited in a number of ways. As examples we will consider two approaches, namely, enhancing product yield and metabolism of xenobiotics.

ENHANCEMENT OF PRODUCT YIELD

In comparison with microbial cultures, plant cells have a slower rate of growth and metabolism and therefore the biotransformation of a precursor to a desirable product may require several days if not weeks. In cases when the conversion reaction in question is confined exclusively to plants and has a high market value, the enhancement of its yield will be of definite industrial interest.

An obvious strategy for yield improvement is the selection of cell strains with high potential for a biotransformation reaction. To avoid confusion a distinction must be made between high yielding plants and high yielding cell cultures.

As the potential to produce secondary metabolites varies in whole plants of a species or even in plants belonging to the same cultivar, it would appear reasonable to select a high-yielding plant for the initiation of cell cultures. In some cases this approach has produced encouraging results. For example, when *Catharanthus roseus* plants with high yields of alkaloids were used to initiate cell cultures, the resulting suspension cultures also had enhanced yields of alkaloid production. However, in other cases a direct correlation between high-yielding parent plants and high-yielding cell cultures was not obvious. When plants of *Digitalis lanata* with identical digoxin contents were used to set up callus cultures, the capacity of the latter to biotransform digitoxin to digoxin (Fig. 7.1) ranged from high rates to almost complete absence of the desired reaction (see later). This phenomenon has been ascribed to the process of transition from organized, highly differentiated plant to the unorganized, undifferentiated cells in culture. The transition may have resulted in mutations and/or chromosomal rearrangements which ultimately impaired the potential for biotransformation and biosynthesis.

Even in cases when high-yielding individual plants give rise to high-yielding cell cultures a variation in the productivity of the latter is not uncommon. The second stage of selection should therefore take into consideration the biosynthetic potential of different cell lines. This evaluation should be undertaken after cell

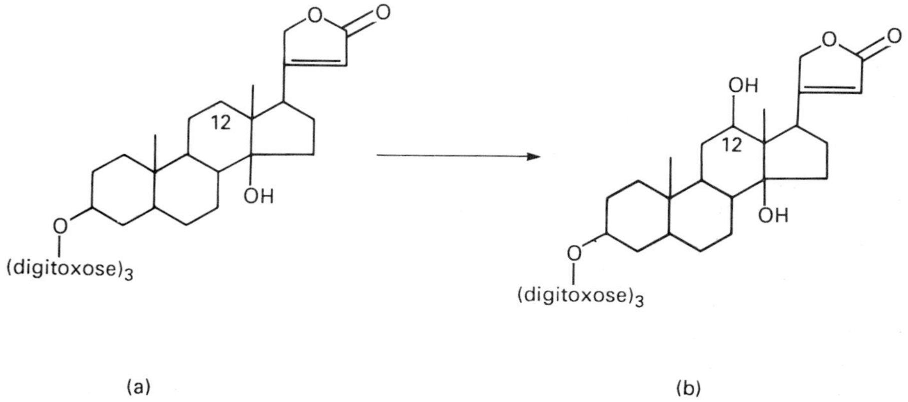

(a) (b)

Fig. 7.1 Structures of (a) digitoxin and (b) digoxin.

suspension cultures have become established; in other words, when they exhibit a certain degree of uniformity in their growth patterns and biosynthetic potential.

Productivity of cell suspension cultures may also be enhanced by a careful control of their physical environment (temperature, lighting), gas regime (e.g. rate of shaking), nutrient medium, and the supply of growth regulators (phytohormones). A clear distinction ought to be made between the enhancement of biomass productivity and that of secondary product synthesis. Examples in which maximum biomass yield coincides with maximum product formation are rare. In general, conditions which favour growth in terms of biomass production have little—and in some cases adverse—effect on secondary metabolite formation. This 'uncoupling' between growth and product formation has led to the formulation of 'production' media which are targeted towards the enhancement of product yield, rather than cell division and growth.

On the basis of this distinction between growth and metabolite productivity, two-stage strategies have been formulated by a number of workers. In the first stage the process is targeted mainly towards the production of cells as rapidly and as cheaply as possible. The biomass thus obtained is then used as the catalytic material for biotransformation in the second stage.

The first stage is carried out by growing cells in batch, semi-continuous, and continuous culture. Batch systems have been extensively used for the cultivation of plant cell suspensions. The cells grow in a closed system in which the amount of nutrients become limiting as the growth progresses and where there is a continuous change in the culture environment and in key parameters such as growth rate. A variation of batch culture is the semi-continuous system in which at the point of maximum biomass, when a certain proportion of cells are still capable of division, about two thirds of the culture is harvested and replaced by fresh medium. In contrast to these methods, the continuous culture constitutes an open system. In this system, the volume of the cultures is kept constant by inflow of a known volume of fresh medium and an equal volume of outflow consisting of cells and spent medium. A steady state can eventually be achieved in the culture, where the generation time, growth rate, and metabolic events in cells remain constant. Continuous cultures do not appear to be of direct use for the production of secondary metabolites. Work with cell cultures of *Galium mollugo* and *Dioscorea deltoidea* has indicated that in continuous cultures the production of respective secondary metabolites, namely anthraquinones and steroids, was repressed significantly.

As mentioned earlier the cell biomass produced through the application of batch, semi-continuous, and continuous systems may be used in a second stage targeted exclusively towards enhancing the yield of biotransformation (Fig. 7.2). The merits of the process are demonstrated in the biotransformation of β-methyldigitoxin by cells of *Digitalis lanata* (Fig. 7.3). The conditions of cell growth required to achieve a high biomass were optimized in a suitable bioreactor. As soon as the desired cell density was achieved in the vessel the cells were harvested and subsequently transferred into a smaller fermenter in which the conditions were optimized for the process of biotransformation. As soon as the cell density in the first fermentor reached the desired level again, the culture was harvested and a

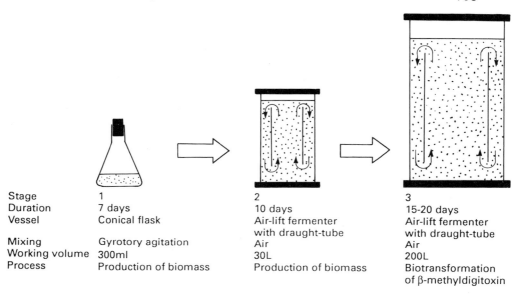

Stage	1	2	3
Duration	7 days	10 days	15-20 days
Vessel	Conical flask	Air-lift fermenter with draught-tube	Air-lift fermenter with draught-tube
Mixing	Gyrotory agitation	Air	Air
Working volume	300ml	30L	200L
Process	Production of biomass	Production of biomass	Biotransformation of β-methyldigitoxin

Fig. 7.2 Scale-up of *Digitalis lanata* cell cultures from 300 mL to 200 L. Transformation of β-methyldigitoxin to β-methyldigoxin.

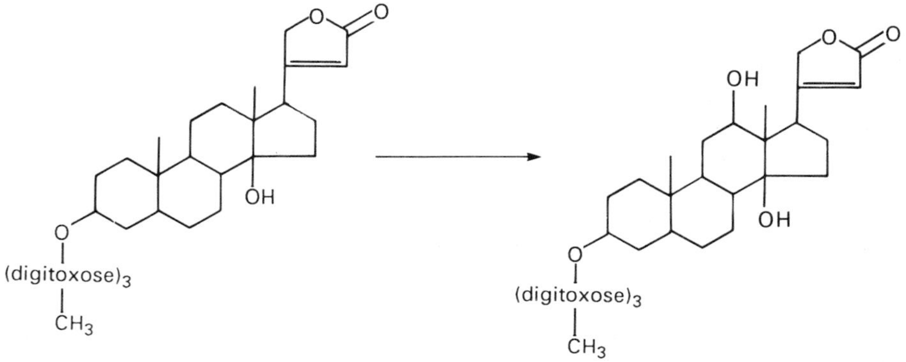

(a)

(b)

Fig. 7.3 Structures of (a) β-methyldigitoxin and (b) β-methyldigoxin.

second biotransformation vessel set up. As a result of setting up cell strains with enhanced hydroxylating capacity and using the described two-stage strategy, Reinhard and Alfermann (1980) and their colleagues were able to increase the yield of product formation through biotransformation by as much as 60-fold.

METABOLISM OF XENOBIOTICS

Another area in which the biochemical potential of plant cell suspensions can be exploited is the degradation of xenobiotics. These chemical compounds, which include pesticides and polycyclic and polychlorinated hydrocarbons, are released into the environment every year in increasing amounts. The present knowledge regarding the degradation of these compounds by plants and the toxicological and other properties of metabolites arising from their degradation is rather limited. The role of plants as animal feedstocks and their human nutritional importance are two factors which make the study of metabolism of xenobiotics by plants an area of special interest. Of course the degradation of xenobiotics through biotransformation by plant cell cultures cannot be regarded as a process which will use inexpensive precursors in order to produce valuable products. Rather it should be seen in the context of the advantages that biotransformation of xenobiotics by plant cell cultures may have in the study of their metabolism by whole plants. In the field it is difficult to dissociate the degradation of xenobiotics by plant and soil microflora and by environmental factors (heat, light) from exclusively plant degradative reactions. For a better understanding of xenobiotic metabolism and the properties of their breakdown products, plant cell cultures provide a practical and advantageous system not readily available in field-grown plants. Furthermore, investigations on the pattern of breakdown of pesticides may provide useful information which will help towards designing new pesticides.

Despite the potential importance of a comprehensive knowledge on the metabolism of pesticides in plants, this field of study is still in its early stages. Initial work on the action of selected plant cell cultures on certain pesticides indicated that cytochrome P-450-dependent microsomal systems, not unlike those present in the liver, may be involved. A discussion of details of the mechanism of action is beyond the scope of this chapter, but interested readers can consult the review article by Sandermann and colleagues (1977).

The metabolic fate of several pesticides in plant cell cultures is illustrated in Table 7.2. Xenobiotics appear to be transformed or degraded by enzymes with broad specificities or specificities for certain functional groups (e.g. peroxidases, hydrolases, D-glucosyl transferases). A significant potential in plant cell cultures—which is absent in micro-organisms—is to conjugate exogenous substances. The most common conjugating moiety encountered in the metabolism of xenobiotics appears to be glucose in the form of glucose ester and glucoside. The most prevalent products of xenobiotic breakdown are β-D-glucosides. As in the case of cardiac glycosides, glycosylation may confer a higher medicinal value to the substrate supplied. Other conjugating moieties involved in xenobiotic metabolism are various amino acids, and also glutathione which is an important intermediary in the elimination of toxic substances in mammalian systems. The conjugation of

Table 7.2 Biotransformation of some xenobiotics by plant cell cultures (from Sandermann *et al.*, 1977)

Substrate	Product	Cell culture
2,4-Dichlorophenoxy acetic acid (2,4-D)		Soybean Tobacco Jackbean Carrot Sunflower Maize
Fluorodifen		Tobacco
Propanil		Rice

exogenously supplied substrates is not a phenomenon confined to pesticides and related compounds. This reaction appears to be common in most cell cultures with biotransformation potential. In consequence, the desired product may not be readily accessible as free species, and this may pose problems at the level of product recovery. This question will be discussed later.

CLASSES OF COMPOUNDS INVOLVED IN BIOTRANSFORMATION REACTIONS

The range of compounds that can serve as substrates for biotransformation reactions by plant cells is almost unlimited. Plant cell cultures have shown potential for the bioconversion of phenylpropanoids (e.g. flavonoids, tannins, anthraquinones) mevalonates (e.g. steroids, cardiac glycosides) and alkaloids. Of course not all these reactions result in products of obvious special industrial (e.g. medicinal) importance. In certain cases they produce compounds which are scarce and sometimes unknown in nature. In others, they reveal the existence of unique reactions within cell culture systems, such as reactions involving optical resolutions of racemic mixtures, and these processes may in future replace presently problematical stages in economically expensive synthetic routes.

Mevalonates

Mevalonates including various terpenoids, steroids and saponins appear to be the most common substrates so far used in biotransformation studies with plant cell cultures. The precursors of terpenoids and steroids are five-carbon units of isopentane and isoprene (Fig. 7.4), which are themselves derived from mevalonic acid, a six-carbon molecule. Saponins, on the other hand are either triterpenoid glycosides or glycosides of steroids with a spiroketal side chain (Fig. 7.5). These include cardiac glycosides (Fig. 7.6) which are derived from the non-acylated, water-soluble form of cholesterol glycosides.

Interest in the biotransformation of cardiac glycosides developed as a result of the need to obtain compounds with efficient cardiac activity and minimum undesirable side effects. Cardiac glycosides are important pharmaceutical agents widely used in the treatment of certain heart disorders. However, as in the case of many drugs, treatment with cardiac glycosides generates a variable toxic effect in patients.

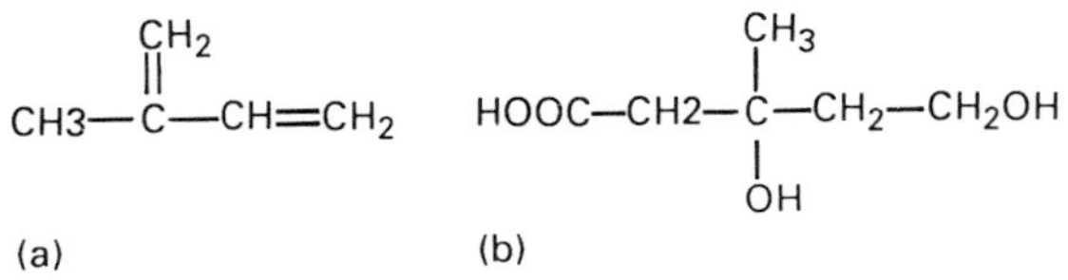

Fig. 7.4   Structures of (a) isoprene and (b) mevalonic acid.

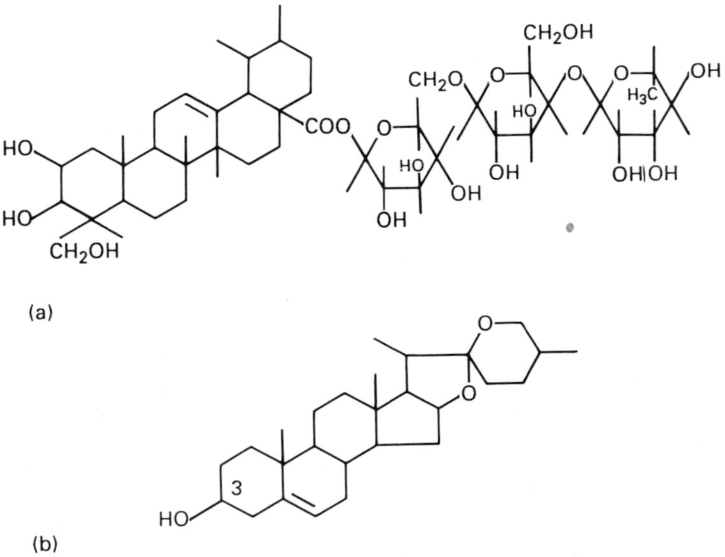

Fig. 7.5   Structures of (a) asiaticoside, a triterpenoid saponin and (b) diosgenin, a steroidal saponin.

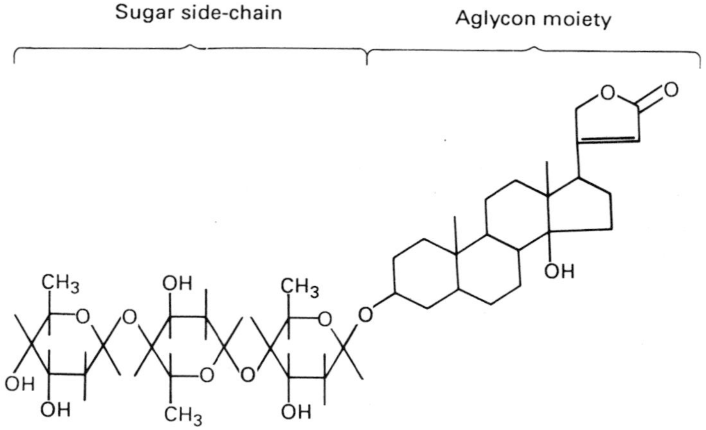

Sugar side-chain   Aglycon moiety

Fig. 7.6 Structure of digitoxin, a cardiac glycoside.

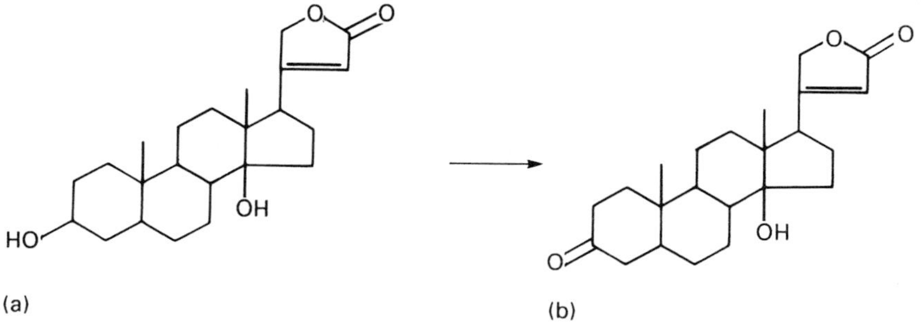

(a)   (b)

Fig. 7.7 Structures of (a) digitoxigenin and (b) 3-dehydrodigitoxigenin.

Early studies directed towards biotransformation of cardenolides were performed with microbial systems. However, due to the low yields obtained attention was focused on cultures of plant cells, which with their inherent genetic variability may produce more diversified products. Screening was undertaken with cell cultures of plant species known to produce cardiac glycosides, namely *Digitalis lanata* and *D. purpurea*. These cultures lacked the metabolic capacity to biosynthesize cardiac glycosides. However, when incubated with compounds of steroidal structure, they were able to bring about potentially useful modifications. Furthermore, the yields of biotransformation products from plant cell cultures were significantly superior to those obtained in microbial systems.

More than two decades ago the first successful studies involving cardiac glycosides were reported by Stohs and Staba (1964). They supplied culture of *D. lanata* and *D. purpurea* with the cardenolide digitoxigenin, and after seven days isolated a product which was later identified as 3-dehydrodigitoxigenin (Fig. 7.7).

Further work in other laboratories indicated that in these cultures the steroid nucleus underwent hydroxylation and glycosylation reactions, and the resulting glycosides were also acetylated.

The hydroxylation potential of plant cell cultures in respect of steroid nucleus is a notable example of the industrial application of these cultures. This is an area in which microbial systems can not compete with plant cells, because of the low efficiency or lack of hydroxylating potential in the former.

Cardiac glycosides consist of a steroid nucleus (aglycon moiety) and a sugar side-chain. Two important examples of cardiac glycosides used in the treatment of heart diseases are digitoxin and digoxin. Digitoxin is one of a class of compounds known as A glycosides. Digoxin, a C glycoside, is derived from digitoxin when the latter is hydroxylated at position C-12 (see Fig. 7.1). In the pharmaceutical industry, both digitoxin and digoxin are extracted from *D. lanata* plants. Due to its special therapeutic properties, digoxin is gradually being used more widely in medicine than digitoxin. However, in the whole plant the amount of digitoxin by far exceeds that of digoxin. As cell cultures of *D. lanata* have been shown to biotransform digitoxin to digoxin, their future potential in the pharmaceutical industry is quite obvious.

Amongst other mevalonates, the biotransformation of steroids has also been the subject of wide-scale study. However, the biotechnological application of the products so far obtained is not as readily apparent as in the case of cardiac glycosides. Furthermore, the biotransformation of certain steroids by plant cell cultures is in most cases similar to their bioconversion by micro-organisms. With plant cell cultures, glycosylation and fatty acid (palmitate) esterification of steroids is more commonly observed than with other organisms. Some of the reactions involved in the biotransformation of steroids include reduction of double bonds or their shift to adjoining positions, stereospecific reduction of keto groups, hydroxylation at a number of sites and oxidation of the 3-hydroxyl group.

The biotransformation of various C_{21} steroids has been extensively studied. These compounds, also known as pregnane derivatives, are degradation products of cholesterol and have widespread occurrence in the plant kingdom. The wide interest in C_{21} steroid metabolism in plant cell cultures can be explained by the fact that some of these steroids occur as precursors in the early stages of cardenolide biosynthesis.

The biotransformation products of progesterone (Fig. 7.8) by a variety of cell cultures are all reported to be 5-α-metabolites in which the hydrogen atom attached to C-5 is orientated below the plane thus conferring a *trans* conformation on rings A and B of the steroid nucleus. No 5-β-metabolites have been reported from plant cell cultures or from cell-free preparations from plant organs, despite the fact that 5-β-derivatives are intermediates in the pathway for cardenolide production from pregnenolone (Fig. 7.9) and progesterone. In contrast, when cell cultures are supplied with 5-β-metabolites of progesterone, they are able to effect stereospecific reductions of 3-keto and 20-keto groups and produce glycoside conjugates through a more or less similar pathway to that of 5-α-pregnane metabolites. However, biotransformations of 5-β-derivatives are much slower than that of 5-α-derivatives. Despite an apparent wide specificity of reactions

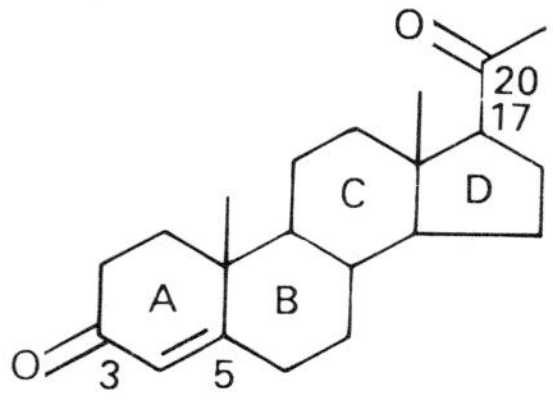

Fig. 7.8 Structure of progesterone. **Fig. 7.9** Structure of pregnenolone.

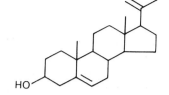

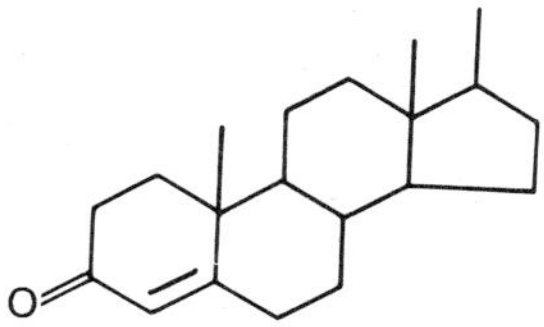

Fig. 7.10 Structure of testosterone.

involved (hydroxylation, reduction, etc.), the enzymes involved in plant cell culture biotransformations exercise discrimination as far as the steric conformation of the molecule is concerned. For example, tobacco suspension cultures were able to metabolize the male sex hormone testosterone (Fig. 7.10) through a range of reactions involving oxidation, reduction, glycosylation, and esterification with the fatty acid palmitic acid. The same cells failed to metabolize related steroids such as cholesterol and sitosterol. This suggests the existence of a steric hindrance as a result of the 8-carbon side chain in the latter substrates and its absence in testosterone.

The presence of a side chain longer than two carbons at C-17 of steroid nucleus in general reduces the efficiency of its bioconversion. Nevertheless, metabolism of cholesterol to diosgenin, an important precursor for various medically important steroid hormones, has been reported from plant cell cultures (Fig. 7.11). The incorporation of sitosterol, the most common sterol in higher plants, into diosgenin has also been reported.

Another class of compounds resulting from the acetate–mevalonate pathway are the terpenoids. Amongst these, the biotransformation of monoterpenoids has attracted most attention in recent years. Monoterpenoids are C-10 compounds resulting from the condensation of two phosphorylated C-5 units. They are volatile substances responsible for much of the characteristic aromas of plants. Monoterpenoids comprise essential oils which play an important role in food (flavouring), cosmetics (perfume), and pharmaceutical industries. Originally considered (like most secondary metabolites) as waste products of metabolism, monoterpenoids in plants act as insect repellents or attractants depending on whether the insects are invaders or pollinators, respectively. In general, undifferentiated cell cultures do not accumulate monoterpenoids to significant levels.

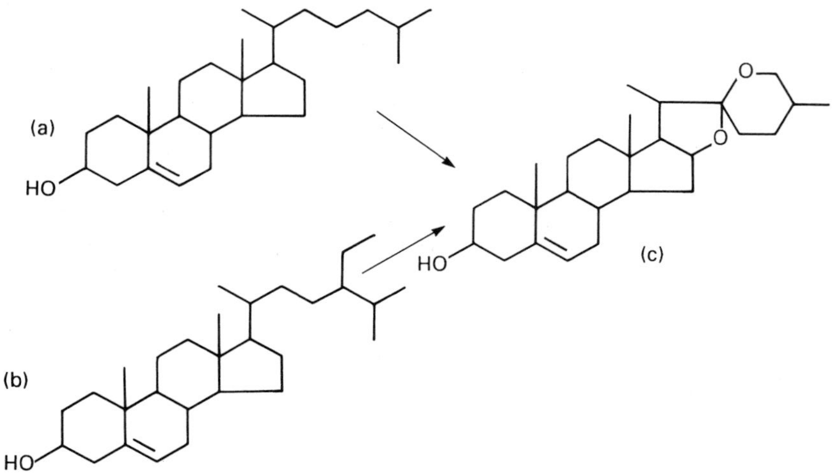

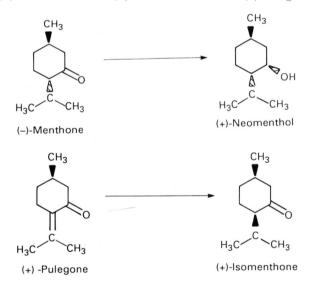

Fig. 7.11 (a) Cholesterol and (b) sitosterol can form (c) diosgenin.

Fig. 7.12 Transformations of monoterpenoids in *Mentha* cell cultures.

This could be due either to the absence of specialized tissues (oil glands) or to rapid breakdown of monoterpenoids following their synthesis in culture. However, following biotransformation studies it is now recognized that cell cultures retain the ability to catalyse at least certain reactions of monoterpenoid biosynthesis pathways.

Cell cultures of different *Mentha* chemotypes have been shown to transform (−)-menthone to (+)-neomenthol by reducing the former's carbonyl group to a hydroxyl group (Fig. 7.12). Both compounds are constituents of peppermint oil and are derived from piperitenone, the precursor of peppermint monoterpenoids. Another constituent of peppermint oil, (+)-pulegone, is bioconverted to (+)-isomenthone through the reduction of its unsaturated double bond. However, the occurrence of this reaction in *Mentha* cell lines is more restricted than the reduction of carbonyl to hydroxyl group. Another example of monoterpenoid biotransformation is the reduction of acyclic and cyclic monoterpenoid aldehydes to their respective primary alcohols. Suspension cultures of *Lavandula angustifolia* devoid of monoterpenoids or their glycosides were able to reduce exogenously added aldehydes geranial, neral, citronellal, and perillaldehyde to their corresponding alcohols geraniol, nerol, citronellol, and perillyl alcohol respectively (Fig. 7.13).

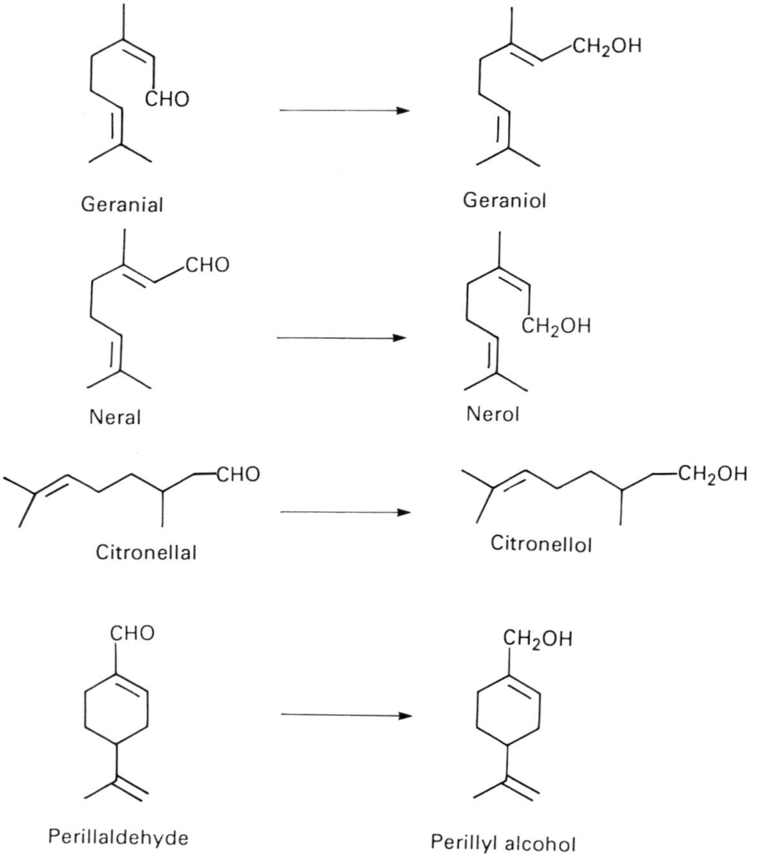

Fig. 7.13 Reduction of aldehydes to primary alcohols by cultures of *Lavandula angustifolia*.

The bioconversions of these compounds were complete in a relatively short period of time, ranging from 15–25 min for the acyclic monoterpenoids geranial, neral, and citronellal to 75 min for the monocyclic monoterpenoid perillaldehyde. The significant difference observed in the rates of these bioconversions could be ascribed to the operation of a functional group-specific reductase with a higher affinity for acyclic rather than cyclic compounds. However, not all enzymes involved in the biotransformation of terpenoids are semi-specific or non-specific. Indirect evidence from cell suspension cultures of *Catharanthus roseus* effecting the bioconversion of the monoterpenoid 10-hydroxygeraniol (Fig. 7.14) suggests the presence of substrate specific enzymes which would reduce the two double bonds of the substrate separately. When 10-hydroxygeraniol was added to a suspension culture of *C. roseus* cells, the first reduced derivatives appeared after 6 h. However, a difference was observed in the rate at which the two double bonds of 10-hydroxygeraniol were reduced. The reaction responsible for the reduction of the 2,3-double bond proceeded at a higher rate than that for the 7,8-double bond (Fig. 7.15). However, both monoreduced products underwent further reduction so that after 48 h the only species present was the completely reduced derivative.

Diterpenoids are formed by the addition of an isoprene unit to a sesquiterpenoid, a class of terpenoids consisting of three isoprene units. The general precursor for diterpenoids is considered to be geranylgeranyl pyrophosphate which is formed

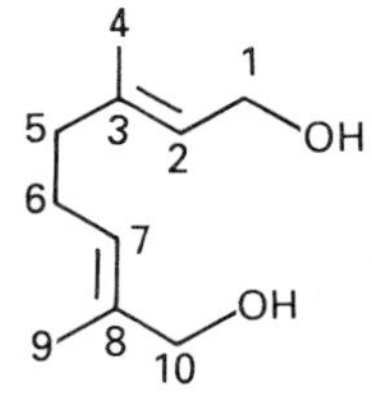

Fig. 7.14   Monoterpenoid structure of 10-hydroxygeraniol.

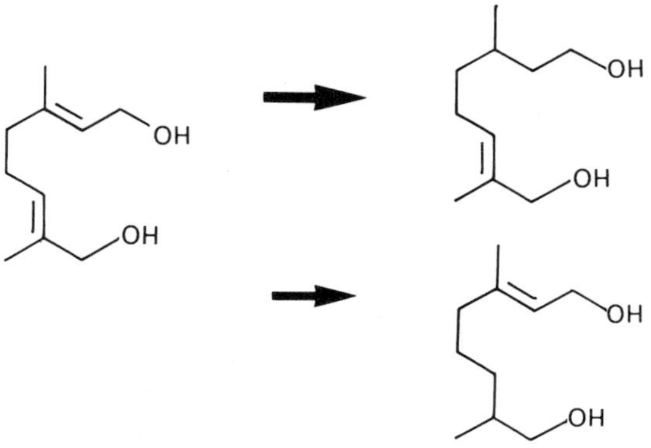

Fig. 7.15   Reduction of 10-hydroxygeraniol.

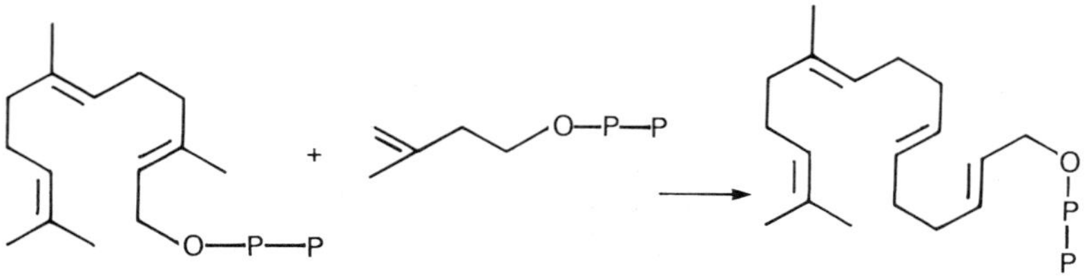

| Farnesyl pyrophosphate | Isopentyl pyrophosphate | Geranylgeranyl pyrophosphate |

Fig. 7.16 Addition of (a) farnesyl phosphate to (b) isopentyl pyrophosphate forms (c) geranylgeranyl pyrophosphate.

(a)   (b)

(c)

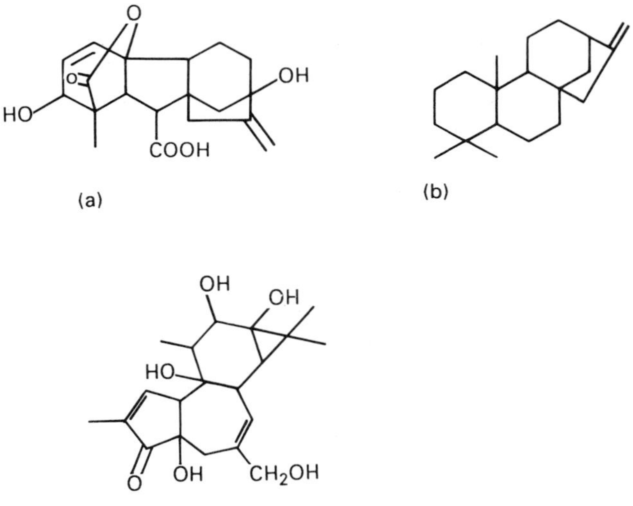

Fig. 7.17 Structures of (a) gibberellic acid, (b) kaurene, (c) phorbol.

by the addition of isopentenyl pyrophosphate to farnesyl pyrophosphate (Fig. 7.16).

The great majority of diterpenoids are C_{20} compounds. Unlike monoterpenoids, diterpenoids are not volatile but occur mostly in resins of higher plants and in fungi. The geometry of the diterpenoid skeleton affords many possibilities for molecular cyclizations. Consequently the number of different diterpenoids found in higher plants and fungi is very large. Most diterpenoids are cyclic compounds such as the plant growth hormone gibberellic acid, one of its precursors kaurene, and phorbol (Fig. 7.17). Alongside the polycyclic diterpenoids there are acyclic types, although these are fewer in number. An important example of an acyclic

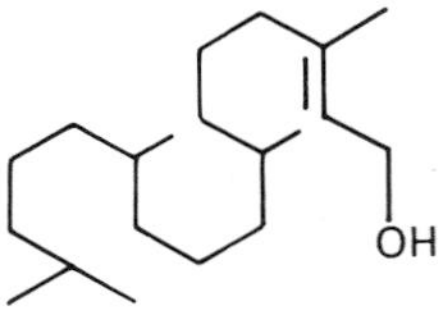

Fig. 7.18   Structure of phytol.

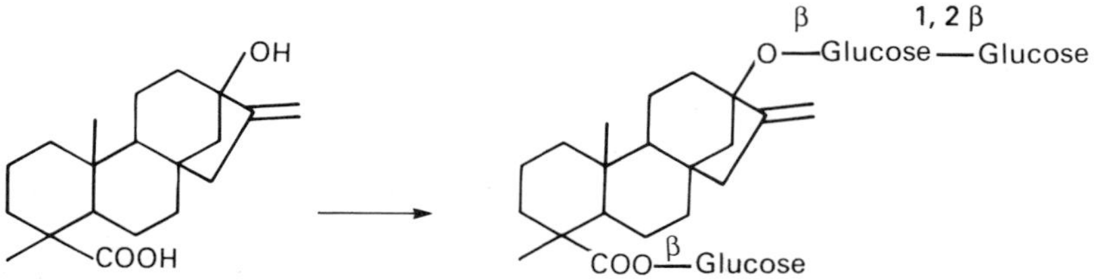

(a)   (b)

Fig. 7.19   Glucosylation of (a) steviol to (b) stevioside.

diterpenoid is phytol (Fig. 7.18) which is a constituent of the chlorophyll molecule as well as being a precursor of the tocopherols (Vitamin E).

An interesting example of diterpenoid biotransformation is the glucosylation of steviol to stevioside by plant cell cultures (Fig. 7.19). The diterpenoid glucoside stevioside was first isolated from the leaves of *Stevia rebaudiana* Bertoni by the French chemists Bridel and Lavieille (1931). *S. rebaudiana*, first described by the French botanist Bertoni, is a perennial shrub belonging to the Compositae. The natural habitat of the plant is in Paraguay and south-west Brazil. The plant leaves have a sweet taste because of the abundant presence of stevioside. This glucoside is about 300 times more sweet than the household sugar sucrose, and consequently is now widely used as a natural sweetener. When cell suspensions of *S. rebaudiana* and *D. purpurea* are supplied with the aglycone steviol, which is not sweet, they glycosylate it to stevioside.

In recent years, stevioside has assumed an increasing importance in countries such as Japan and Brazil as a natural substitute not only for sucrose but also for other non-calorific sweeteners. The economic importance of stevioside will be considerably enhanced if the use of artificial sweeteners such as saccharin is discouraged or prohibited worldwide on medical grounds. Another product of steviol glucosylation is rebaudioside A (Fig. 7.20) which contains one more glucose moiety than steviosole. Although available in the whole plant in lower yield than stevioside, this glucoside is reported to have superior taste properties. There are as yet no reports of rebaudioside A as a product of steviol biotransformation. However, it is not impossible that future investigations in this area would lead to rebaudioside A as well as other glucosides with enhanced taste characteristics.

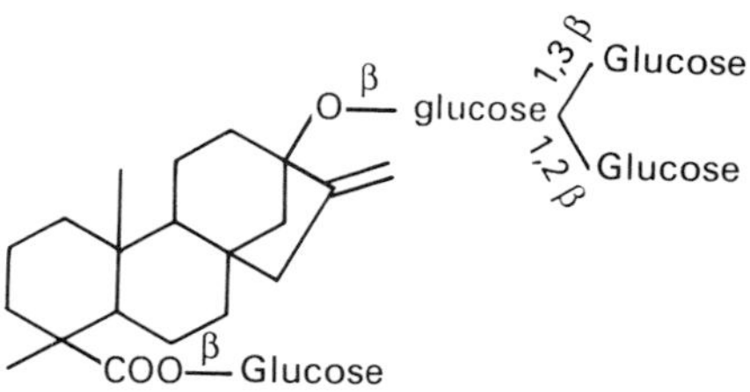

Fig. 7.20 Structure of rebaudioside A.

Phenylpropanoids

Phenylpropanoids are C_6–C_3 residues (Fig. 7.21) involved in the biosynthesis of a large number of plant secondary products. These products, almost all of which are aromatic, arise from different origins. They are either derived from the shikimic acid pathway, for example phenols, cinnamic acid derivatives, and coumarins, or possess a mixed biogenesis, i.e. their skeletons are derived from at least two different pathways. Most compounds in the second category are coloured. They include flavonoids such as anthocyanins, quinones such as anthraquinone and xanthones.

In higher plants the abundance of phenolic substances is second only to carbohydrates. The association of plant phenolics with human culture, as pigments, tanning agents, and medicinals has a long history. The structural variety of phenolics in plants is as wide as their occurrence; they range from the relatively simple molecule of salicylic acid (Fig. 7.22) to the complex polymeric structure of lignin.

Simple phenols are monocyclic aromatic compounds arising from the deamination of phenylalanine and tyrosine. Phenolic substances were among the first to be investigated in biotransformation studies with plant cell cultures in the mid-1970s. The choice was dictated by the need to gain a better understanding of glucosylation processes in plant cells, especially after their potential for the production of useful glucosides (e.g. from certain steroids and cardenolides) became known.

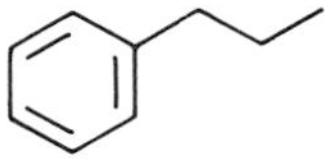

Fig. 7.21 Phenylpropanoid configuration.

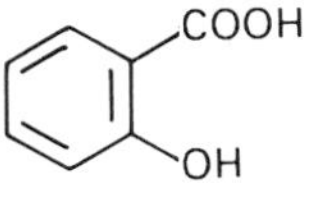

Fig. 7.22 Structure of salicylic acid.

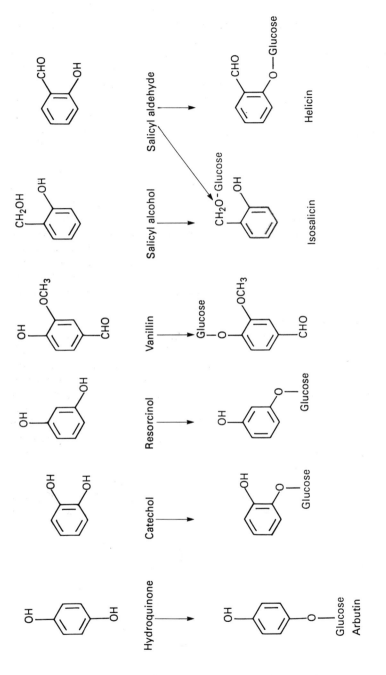

Fig. 7.23 Glycosylation of phenols.

One of the earliest biotransformations studied involving phenols was under-taken with callus cultures of *Argostemma githago*, *Datura ferox*, and *Digitalis purpurea*. Compounds with different hydroxylation patterns were supplied in the nutrient medium and the products were analysed at the end of the growth cycle. Callus cultures of all three species were able to transform the exogenously supplied substrates, with *D. purpurea* tissue showing less potential than the other two species. The substrates were catechol, resorcinol, vanillin, salicyl alcohol, and salicyl aldehyde, and the biotransformation reaction involved was glycosylation (Fig. 7.23). Both salicyl alcohol and salicyl aldehyde were converted to isosalicin, with only traces of the expected product from the glycosylation of salicyl aldehyde, namely helicin, detected. These early results were supported by later work with cell suspension cultures of *Datura innoxia*. Although both resorcinol (*m*-dihydroxybenzene) and catechol (*o*-dihydroxybenzene) yielded their respective monoglucosides, the rate of hydroquinone biotransformation was quite remark-able. Within 10 h after the addition of hydroquinone to the cell cultures, it was totally converted to its glucoside, arbutin. The same cell cultures converted both salicyl alcohol and salicyl aldehyde to isosalicin rather than to their respective glucosides salicin (Fig. 7.24) and helicin. The story of the glucosylation pattern of salicyl alcohol is an illustrative example of how plant cell suspension cultures in general, and biotransformation studies in particular, may contribute to the elucidation of biosynthetic pathways of secondary metabolites in plants.

The glucoside salicin (*o*-hydroxymethyl phenyl β-D-glucoside) was the first glucoside isolated from natural sources. Used externally, it is an analgesic for rheumatic pains. As such, salicin was the precursor of aspirin (acetylsalicylic acid) with its analgesic and antipyretic properties. Salicin occurs in certain plants of the Rosaceae family, but salicyl alcohol is restricted to willows (*Salix* spp.). *In vivo* and *in vitro* studies with various plant materials (cultured cells and plant organs) in the 1960s and 1970s had shown that the glucoside salicin is derived from *o*-coumaric acid or benzoic acid via the intermediate salicyl aldehyde and its glucoside, helicin (Fig. 7.25). The same experiments also suggested that the glucosylation of salicyl alcohol involves the alcoholic rather than the phenolic group and consequently salicyl alcohol could not be a direct precursor of salicin. This pathway of salicin biosynthesis was widely accepted by the early 1980s and appeared in reviews and textbooks of plant secondary metabolism. However, recently the question has come under review with salicyl alcohol reinstated as a precursor for salicin. A

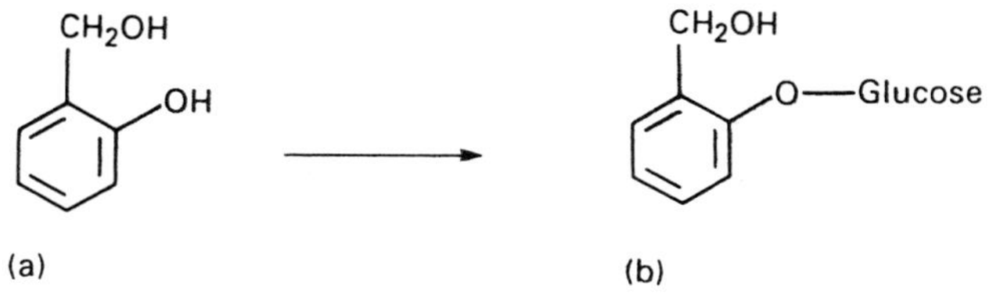

Fig. 7.24 Conversion of (a) salicyl alcohol to (b) salicin.

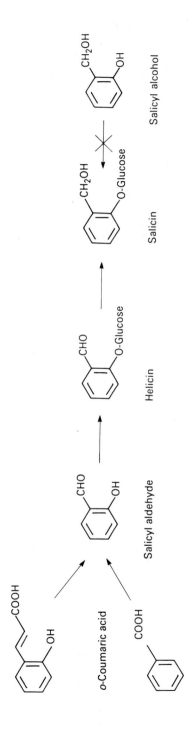

Fig. 7.25 Biosynthesis of salicin.

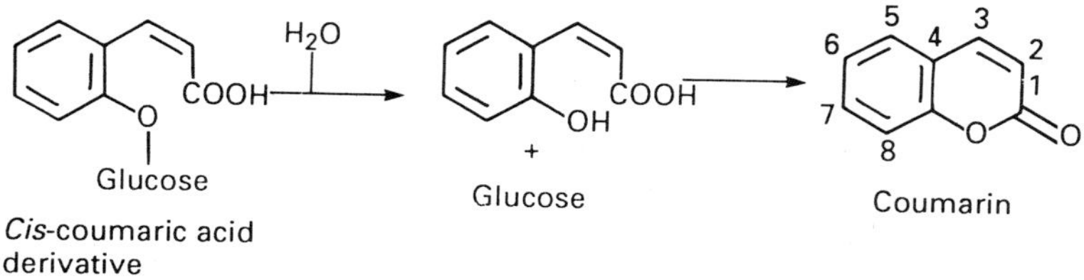

Fig. 7.26 Biosynthesis of coumarin.

Japanese group led by H. Mizukami reported in 1983 that cell suspension cultures of *Lithospermum erythrorhizon* and *Gardenia jasminoides*, in addition to isosalicin production, were also able to glycosylate the phenolic group of salicyl alcohol to produce salicin (see Fig. 7.24). Later the same group isolated and partially characterized an enzyme catalysing this glucosylation reaction in cultured *G. jasminoides* cells (Mizukami *et al.*, 1983). The enzyme, termed UDP-glucose:salicyl alcohol phenyl-glucosyltransferase, showed a clear specificity for the phenolic position of salicyl alcohol. The highest activity in the cells was during the exponential phase of growth when about 70% of the added substrate was glycosylated within four days. The growth stage of cells also appeared to influence the ratio of salicin:isosalicin formation, with the ratio decreasing as the cells reached the stationary phase. Furthermore, the auxins 2,4-D and NAA suppressed salicin formation and increased concentrations of medium sucrose enhanced it. As an example, the investigation of glycosylation of salicyl alcohol has illustrated the usefulness of biotransformation studies with cultured plant cells in elucidating the finer points of plant secondary metabolism pathways and their regulatory mechanisms.

Another class of phenolic compounds used in biotransformation studies are the coumarins. Like most other phenolics, coumarins arise from cyclization of *cis*-coumaric acid derivatives which are less stable than their *trans* isomers. In the case of coumarins, the C_3 portion of C_6–C_3 phenylpropanoid configuration is in the form of an unsaturated lactone ring (Fig. 7.26).

Both synthetic and naturally occurring coumarins have a wide range of uses, from anticoagulants to fluorescent dyes. Coumarin itself is present in volatile substances responsible for the odour of numerous plant species. For example, in sweet clover the characteristic scent of new-mown hay is due to the presence of coumarin. Coumarin, together with other unsaturated lactones, is also an inhibitor of germination in some seeds.

Of the several hundred naturally occurring coumarins, the great majority are oxygenated at position 7. The simplest representative of this group is umbelliferone (Fig. 7.27) which is considered to be the precursor of most coumarins *p*-hydroxylated at position 7. Cell cultures of *Catharanthus roseus* and *Conium maculatum* hydroxylate coumarin at position 7 to yield umbelliferone. The latter, when administered to cell suspensions of garden rue (*Ruta graveolens*) is transformed into demethylsuberosin (Fig. 7.28), which is considered to be the precursor of a class of secondary metabolites known as furanocoumarins (see below).

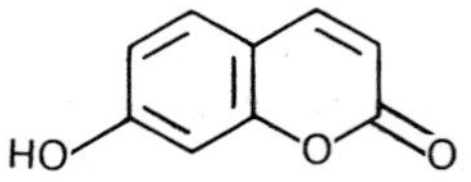

Umbelliferone

Fig. 7.27   Structure of umbelliferone.

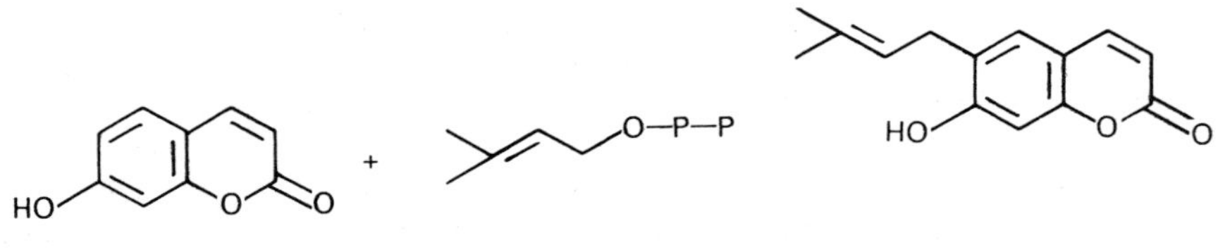

(a)   (b)   (c)

Fig. 7.28   (a) Umbelliferone and (b) dimethylallyl pyrophosphate form (c) demethylsuberosin.

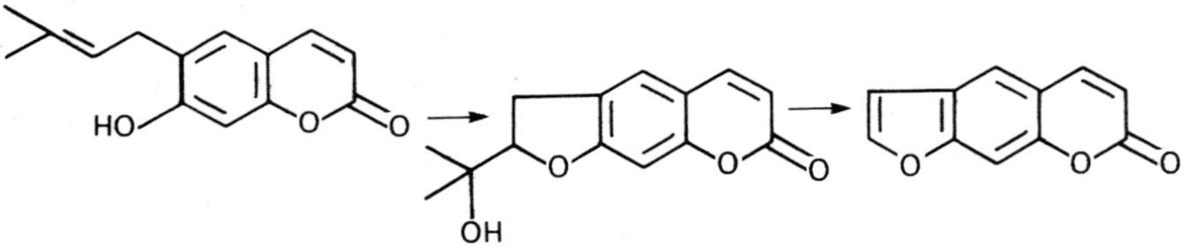

(a)   (b)   (c)

Fig. 7.29   Ring closure in (a) demethylsuberosin yields (b) marmesin and eventually (c) psoralen.

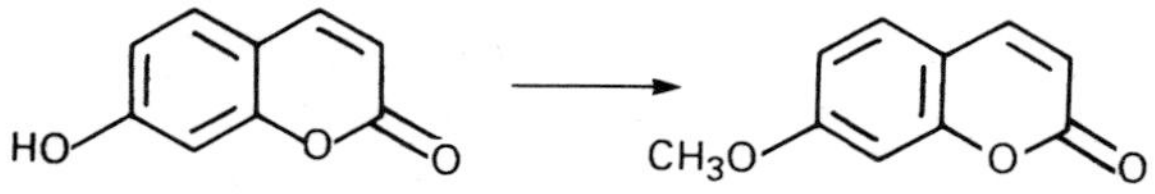

(a)   (b)

Fig. 7.30   Conversion of (a) umbelliferone to (b) herniarin.

The condensation reaction is catalysed by the enzyme prenyltransferase (umbelliferone:dimethylallyltransferase) which has been isolated from the chloroplasts of garden rue cells. The biotransformation of umbelliferone proceeds further by the ring closure in demethylsuberosin to yield the furanocoumarin marmesin and eventually, through degradation of the side chain, psoralen (Fig. 7.29). In a separate reaction, this time involving methylation, umbelliferone is converted to the coumarin derivative herniarin (Fig. 7.30).

Steck and Constabel, who had first studied the above biotransformation sequences in *R. graveolens* cell cultures (1974), also investigated the specificity of the enzymes catalysing them. The cells were supplied with analogues of umbelliferone methylated at positions 4 and 8. The pattern of biotransformation observed for these unnatural substances strictly paralleled that of umbelliferone with the production of 4- and 8-methyl analogues of marmesin and herniarin and possibly 8-methylpsoralen. These results indicated that alongside biotransformations where the substrate specificity is paramount, there exist instances where the specificity of the enzymes involved is directed principally towards a functional group. Such cases offer unlimited possibilities for the biotransformation of compounds of synthetic origin and those obtained from non-plant sources.

Biotransformation studies with plant cell cultures involving coumarins have contributed greatly to our understanding of the process. Another example is the glycosylation of esculetin (6,7-dihydroxycoumarin) to esculin (Fig. 7.31). Cell cultures of *Lithospermum erythrorhizon*, *Gardenia jasminoides*, and tobacco (*Nicotiana tabacum*) were all able to transform the aglycone esculetin to its glucoside esculin, although *L. erythrorhizon* cells were by far the most efficient. Furthermore, among the various cell lines of *L. erythrorhizon* marked differences were observed in glucosylating capacity which ranged from 0.06 µmol glucoside/g fresh weight to 1.87 µmol glucoside/g fresh weight produced 24 h after the aglycone administration. The rate of biotransformation in high-yielding cultures was also significantly influenced by the growth stage and the type of growth regulator used in the nutrient medium. Thus careful selection of nutrient factors, cell lines and growth stages can increase the efficiency of biotransformation processes.

Earlier mention was made of a class of compounds used in biotransformation reactions and known as furanocoumarins. Although these compounds can be viewed as derivatives of coumarins, they are in fact products of mixed biogenesis as their precursors arise from the shikimate pathway (umbelliferone) and the acetate–mevalonate pathway (dimethylallyl pyrophosphate). Biotransformation

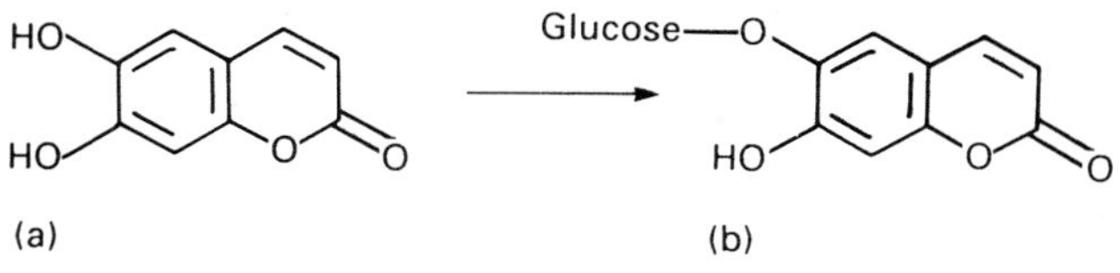

(a) (b)

Fig. 7.31 Glycosylation of (a) esculetin to (b) esculin.

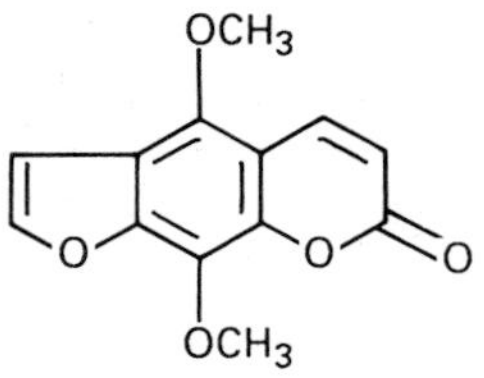

Fig. 7.32 Structure of isopimpinellin.

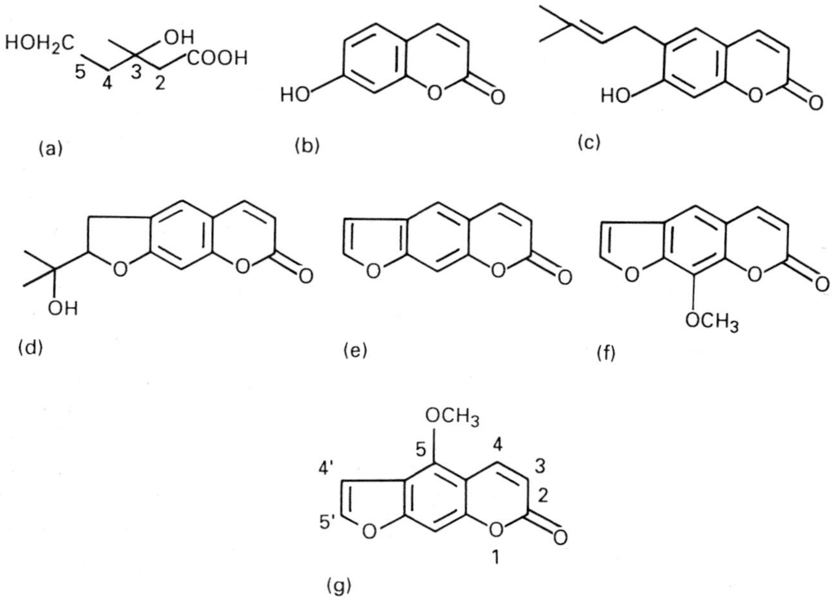

Fig. 7.33 Furanocoumarins and precursors: (a) mevalonic acid, (b) umbelliferone, (c) demethylsuberosin, (d) marmesin, (e) psoralen, (f) xanthotoxin, (g) bergapten.

studies have mainly been used to elucidate the biosynthetic pathway of furanocoumarins and their interconversions. Most of these studies have been carried out with cell cultures of *R. graveolens*, which is a rich source of furanocoumarins. These cultures are able to synthesize a furanocoumarin, isopimpinellin (Fig. 7.32), which is not present in organized *Ruta* plants.

Cell cultures of *R. graveolens* are able to utilize umbelliferone, 7-demethylsuberosin and marmesin as precursors for the furanocoumarins psoralen, xanthotoxin, bergapten, and isopimpinellin (Fig. 7.33). The poor incorporation of mevalonic acid in these experiments raised questions about its biosynthetic role as a precursor of the furan ring. However, other studies with cell cultures of

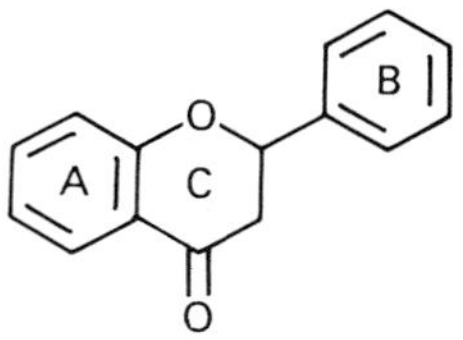

Fig. 7.34 Flavonoid structure.

Thamnosma montana using tritiated mevalonic acid have firmly established that its carbon atoms 4 and 5 become positions 5' and 4' of the furan ring in furanocoumarins.

The synthetic route of the novel compound isopimpinellin was also studied in *R. graveolens* cultures. These cells are able to biotransform both xanthotoxin and bergapten to isopimpinellin through hydroxylation and subsequent methylation reactions. However, the biosynthetic route via xanthotoxins was preferred possibly because of the higher susceptibility of this intermediate to hydroxylation. The class of aromatic compounds collectively known as flavonoids are characterized by a C_6–C_3–C_6 structure in which the C_3 group links the two benzene rings and has different substitution patterns (Fig. 7.34). These patterns in turn define the various flavonoid subgroups. The benzene rings to the left and right of the oxygen heterocycle (ring C) are designated A and B respectively. Flavonoids are widespread throughout the plant kingdom, where they occur mostly as glucosides. The presence of double bonds in flavonoid molecule causes them to absorb visible light and give colour to plant tissues (mainly flowers) in which they are present.

Nearly all biotransformations with flavonoids as substrates result in one-step glycosylations. As such, they have provided valuable information on the influence of the pattern of substitution in the flavonoid skeleton on the specificity of glycosyltransferase enzymes. For example when the three flavonoids liquiritigenin (7,4'-dihydroxyflavanone), naringenin (5,7,4'-trihydroxyflavanone), and baicalein (5,6,7-trihydroxyflavone) (Fig. 7.35) with differing patterns and degrees of substitution were supplied to cell cultures of *Datura innoxia, Perilla frutescens, Catharanthus roseus,* and *Lithospermum erythrorhizon*, the highest yield of glycosylation was observed with the least substituted compound, namely liquiritigenin. Naringenin which has an additional hydroxyl group in position 5 was glycosylated only in *P. fructescens* and *Bupleurum falcatum* cultures with substantially reduced yields. As for baicalein with its three hydroxy substitutions on adjoining positions, no biotransformation was observed, most probably due to steric hindrance. The 2,3-dehydrogenation of naringenin gives rise to the corresponding flavone, apigenin. When administered to suspension cultures of *Cannabis sativa* the latter is *O*-glycosylated at position 7 and *C*-glycosylated at position 8 to yield vitexin (Fig. 7.36). The *O*-glycosylation of flavones with free hydroxyl groups is very common in nature. The *C*-glycosides are much rarer and are mainly apigenin derivatives. The exact mechanism of *C*-glycosylation in the intact plant is not fully understood, although experiments with labelled intermediates have suggested that it may

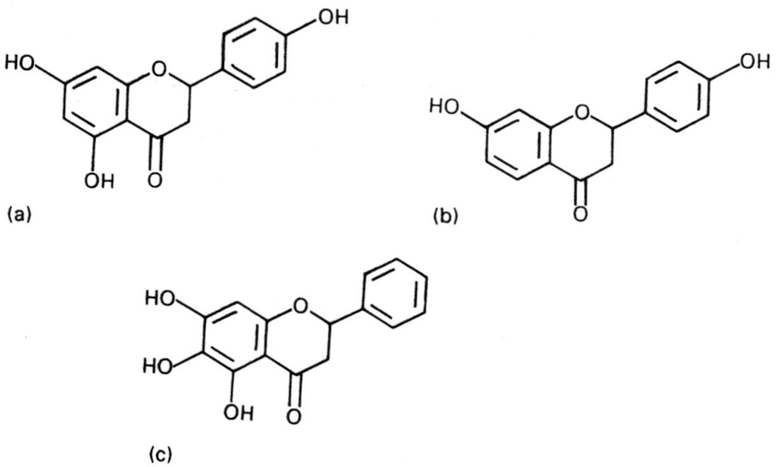

Fig. 7.35   Three flavonoids: (a) naringenin, (b) liquiritigenin, (c) baicalein.

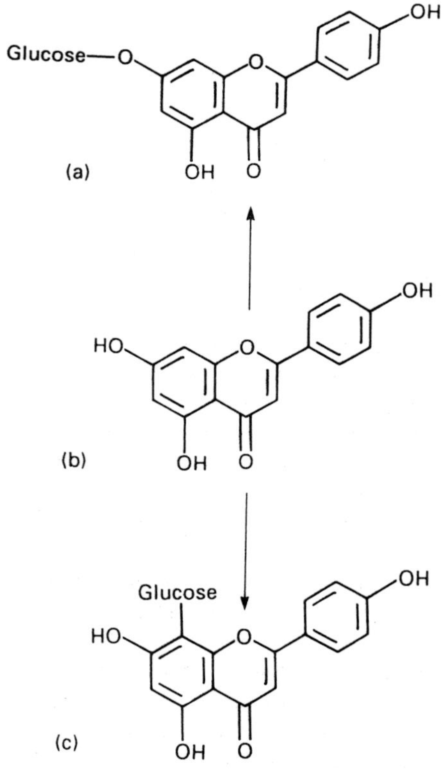

Fig. 7.36   (a) Apigenin-7-*O*-glucoside, (b) apigenin, (c) apigenin-8-*C*-glucoside (vitexin).

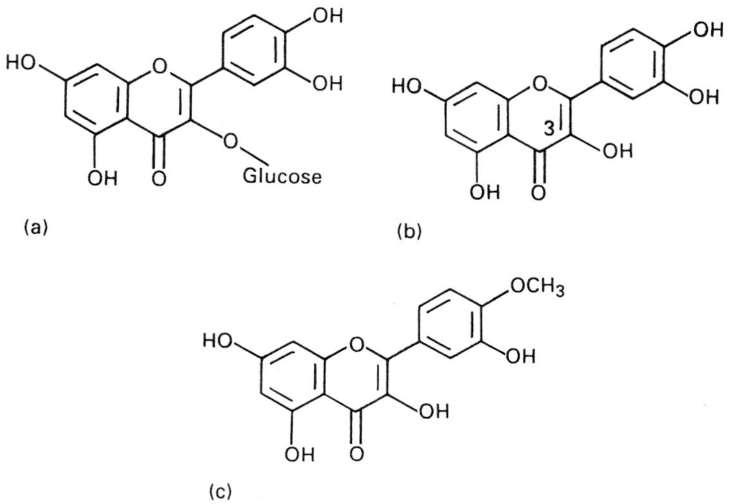

Fig. 7.37 (a) Quercitin-3-*O*-glucoside, (b) quercitin, (c) 3'-*O*-methyl quercitin.

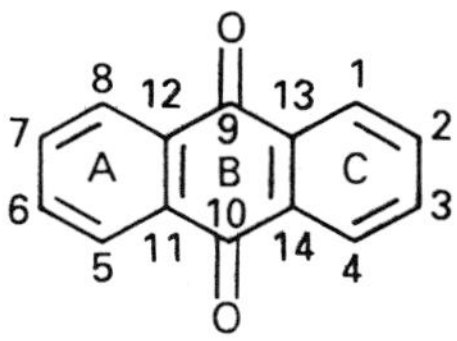

Fig. 7.38 Anthraquinone skeleton.

occur prior to ring closure in flavones. The bioconversion of apigenin with *C*-glycosylation was unknown until it was reported recently by Braemer and his colleagues (1987) working with cell suspensions of *C. sativa*. These cultures were also able to *O*-glycosylate the flavonol quercetin in position 3' (Fig. 7.37). However, they also exhibited a novel biosynthetic capacity for methylation of the hydroxy group on C-3'. As no *O*-methylation from whole plants has so far been reported, this was yet another example of possibilities available with plant cell biotransformations.

Although the biosynthesis of another class of coloured substances, namely anthraquinones (Fig. 7.38), has been studied in several plant species in culture, reports of their biotransformation are scarce. Of the naturally occurring quinones, anthraquinone pigments are the most numerous and most widely distributed in plants. Several anthraquinones have been used as natural dyes as well as drugs because of their cathartic properties. Most anthraquinones occur as glycosides in higher plants. Glycosylation is also the form of biotransformation of these pigments by plant cell cultures. Cells of *Perilla frutescens* glycosylate alizarin (a natural dye from the root of the madder, *Rubia tinctorum*), rhein, and emodin (a pigment with

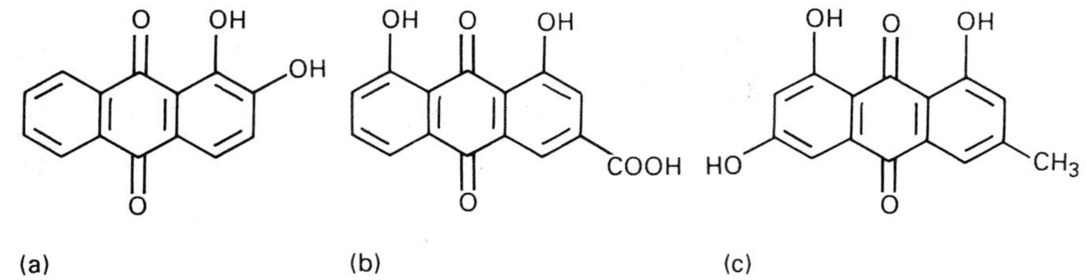

Fig. 7.39   (a) Alizarin, (b) rhein, (c) emodin.

strong cathartic properties) (Fig. 7.39) with decreasing efficiency in that order, probably because of increased substitution in the rings A and C. Rhein is also glycosylated, probably in position 8, by cell cultures of *Catharanthus roseus*, *Lithospermum erythrorhizon*, and *Gardenia jasminoides* with conversion rates ranging from 2.2 to 11.6%. These biotransformation patterns are in accordance with the general trend of conversion of phenylpropanoid derivatives (in particular when glycosylation is involved), namely that the efficiency of biotransformation is higher when the molecules are small and possess fewer substituents in the vicinity of the hydroxy group.

Alkaloids

Among the great variety of secondary metabolites produced by the higher plants, alkaloids occupy a special place. They have remarkable effects on the physiology of animals, and throughout history have been used in numerous ways mostly as medicines. Nearly 6000 different alkaloids are now known. They exhibit a great variety of structural types with very little identity or unity as a group. The only common factors among this diverse collection of organic compounds are that they have cyclic structures and contain nitrogen as part of a heterocyclic ring system. The presence of nitrogen confers basic properties to alkaloids, an early observation which gave the group its name. Although widespread in the plant kingdom, alkaloids are more often found in the dicotyledons.

Because of their uses in orthodox medicine and their commercial importance alkaloids have received a lot of enthusiastic scientific attention. Morphine, found in the opium poppy (*Papaver somniferum*) over 150 years ago, was the first alkaloid to be isolated and crystallized. Over many years, organic chemists have isolated and characterized the alkaloids, their precursors, and some intermediates of their biosynthetic pathways. However, most enzymatic steps responsible for the formation of alkaloids still await elucidation and the information available on aspects of their metabolic regulation is meagre. The advent of plant cell culture was seen as an ideal alternative for the study of plant secondary metabolism, including that of alkaloids. However, these expectations were moderated when it was realized that not all these pathways are expressed in full in undifferentiated plant systems. Despite these setbacks, the potential of non-producing cultures to perform attractive single-step biotransformations was soon realized and has since been applied in alkaloid bioconversion studies.

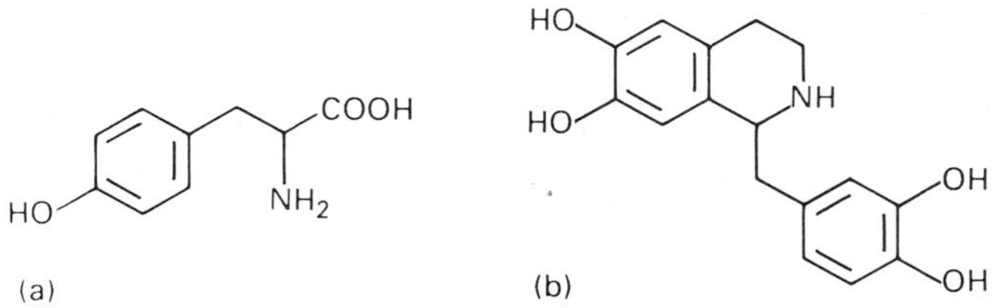

Fig. 7.40   (a) Tyrosine, (b) norlaudanosoline.

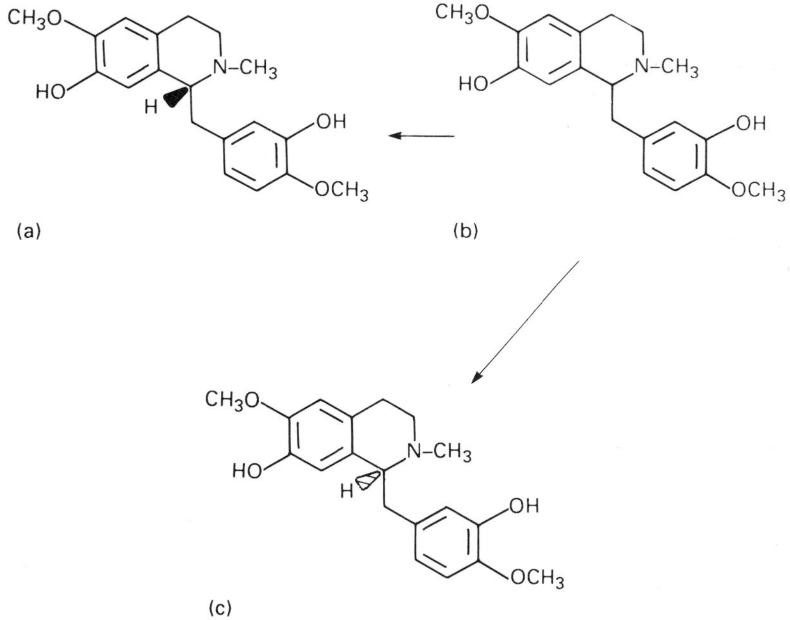

Fig. 7.41   (a) $(-)$-(R)-Reticuline, (b) (RS)-reticuline, (c) $(+)$-(S)-reticuline.

Interesting biotransformation work has been carried out with a class of alkaloids known as benzylisoquinoline alkaloids. These compounds which are important pharmacologically are derived from two molecules of the aromatic amino acid tyrosine and have norlaudanosoline as their immediate precursor (Fig. 7.40). Morphine and codeine are the best known benzylisoquinoline alkaloids. Despite scattered reports of the presence of those alkaloids in cultured cells, the overwhelming majority of undifferentiated *P. somniferum* cultures have not been found to accumulate the morphinan alkaloids. An important intermediate of the biosynthetic pathway after norlaudanosoline is reticuline. This compound is a focal intermediate in the synthesis of morphinan and berberine alkaloids with $(-)$-(R)-reticuline and $(+)$-(S)-reticuline as the immediate precursors of their respective pathways (Fig. 7.41). In contrast to alkaloids derived from $(-)$-(R)-

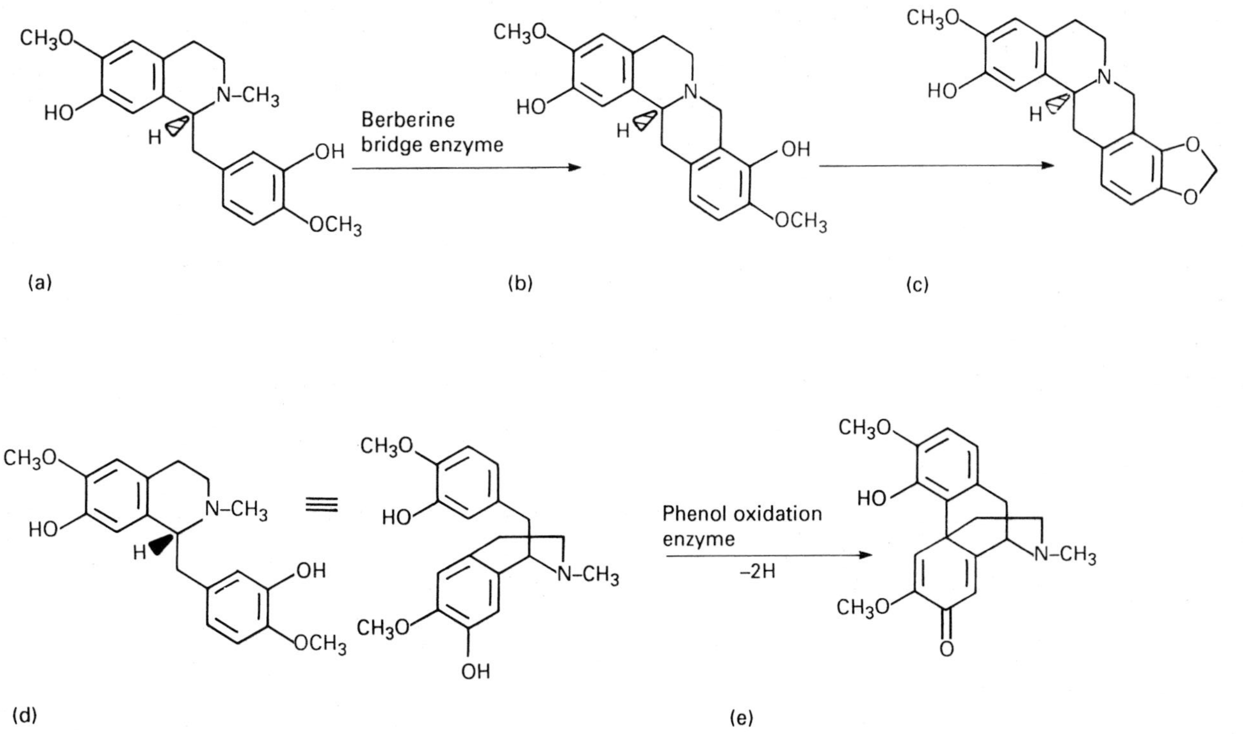

Fig. 7.42 (a) (+)-(S)-Reticuline, (b) (−)-(S)-scoulerine, (c) (−)-(S)-cheilanthifoline, (d) (−)-(R)-reticuline, (e) (+)-salutaridine.

reticuline, the presence of those originating from $(+)$-(S)-reticuline in tissue and cell cultures of *P. somniferum* is well documented. In order to gain a better understanding of the metabolic fate of reticuline, the biotransformation of a racemic mixture of reticuline [(RS)-reticuline] by *P. somniferum* cells was investigated by Furuya and his colleagues (1978). Their results provided interesting information on the expression and repression of secondary pathways in plant cells. The $(+)$-(S)-reticuline of the racemic mixture was selectively and stereospecifically cyclized to $(-)$-(S)-scoulerine, with a biotransformation ratio of 14%, and to $(-)$-(S)-cheilanthifoline with a reduced ratio of less than 1%. In contrast the stereoisomer $(-)$-(R)-reticuline was not metabolized and was recovered unchanged. The biotransformation of reticuline to scoulerine was catalysed by the so-called 'berberine bridge' enzyme, whereas the failure to metabolize $(-)$-(R)-reticuline was due to the absence of the phenol oxidation enzyme (Fig. 7.42).

As both stereoisomers are known to be incorporated into morphine in the whole plant, the result obtained with cell cultures would seem to point also to the absence in undifferentiated tissue of a racemase-type enzyme catalysing the interconversion of reticuline isomers. The failure of the appropriate reticuline stereoisomer to enter the morphinan biosynthetic pathway would have pointed to the possibility of the absence of the subsequent biosynthetic steps in *P. somniferum* cell cultures. Interestingly, however, poppy cell cultures are able to biotransform certain intermediates of the later stages of morphine biosynthesis. In the whole plant of *P. somniferum* the biosynthesis of morphine from salutaridine proceeds with the closing of one more ring to yield thebaine. Demethylation of the latter gives rise to neopinone and subsequently codeinone which is then reduced to codeine. A final *O*-demethylation of codeine yields morphine (Fig. 7.43). Cell cultures shown to be incapable both of metabolizing $(-)$-(R)-reticuline and of biosynthesizing morphine were nevertheless able to biotransform supplied codeine in three days with a biotransformation ratio of 61%. This pattern of biotransformation was in accordance with that observed in whole plants of *P. somniferum*. However, when thebaine was supplied to cultured cells, neopine (Fig. 7.44) rather than neopinone, was found to be the major product with a conversion ratio of 4% in three days. Neopine is a reduced form of neopinone and has not been reported from the poppy plant or postulated in the biosynthetic pathway of morphinan alkaloids. Thus the intermediates of the biosynthetic pathway of a given natural product in undifferentiated systems may differ from those postulated or established for the whole plant. Although such unexpected 'deviations' may be partly responsible for the absence of a desired natural product in cell cultures, they nevertheless provide infinite possibilities for biotransformation studies.

Papaverine (Fig. 7.45) is another isoquinoline alkaloid which has been used recently in a number of biotransformation studies. An opium alkaloid, papaverine is used as a vasodilator drug acting on blood vessels through relaxation of smooth muscle. In the whole plant (the opium poppy) it is formed from norlaudanosoline and $(-)$-norreticuline. The free hydroxyl groups of the latter are methylated before the nitrogen is dehydrogenated to yield papaverine (Fig. 7.46). A variety of biotransformations is performed by suspension cultures of

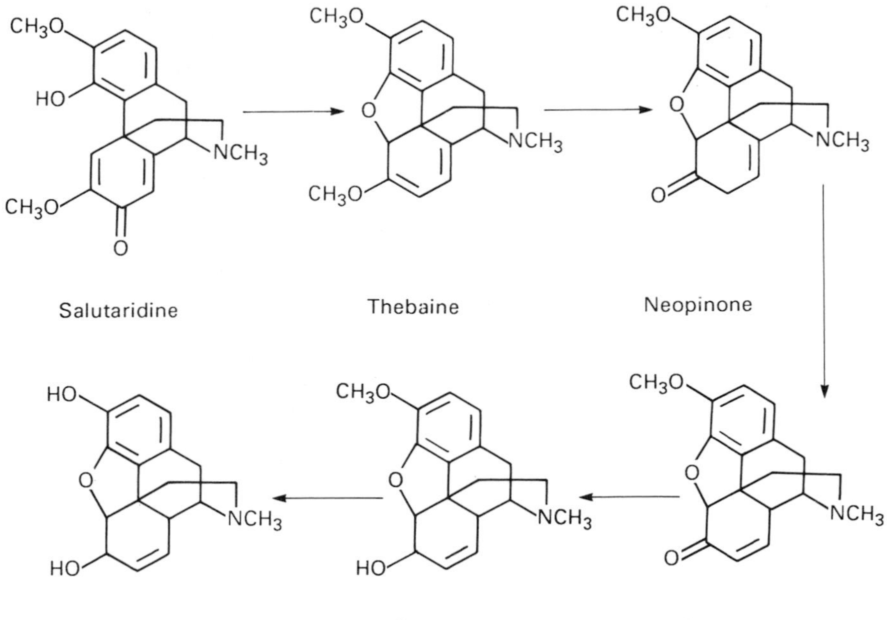

| Salutaridine | Thebaine | Neopinone |

| Morphine | Codeine | Codeinone |

Fig. 7.43 Biosynthesis of morphine.

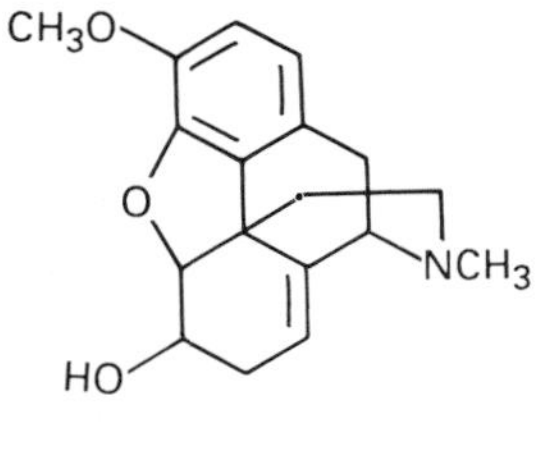

Neopine

Papaverine

Fig. 7.44 Structure of neopine. 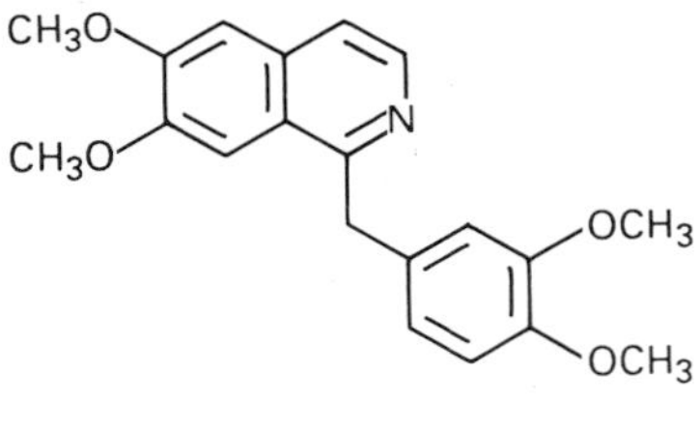   **Fig. 7.45** Structure of papaverine.

Silene alba cells on papaverine. The cultures are suitable systems for biotransformation studies because *S. alba* cells do not produce alkaloids and consequently the separation and identification of exogenous substances and products of their possible bioconversion present fewer problems. The transformation of papaverine in these cultures proceeds via two regioselective *O*-demethylations to yield 6-monodemethylpapaverine and 4′-monodemethylpapaverine (Fig. 7.47). The appearance of these two products was observed at different times during the culture cycle suggesting a dependence between growth stage and the potential for specific demethylation. Whereas most of 4′-monodemethylpapaverine was pro-

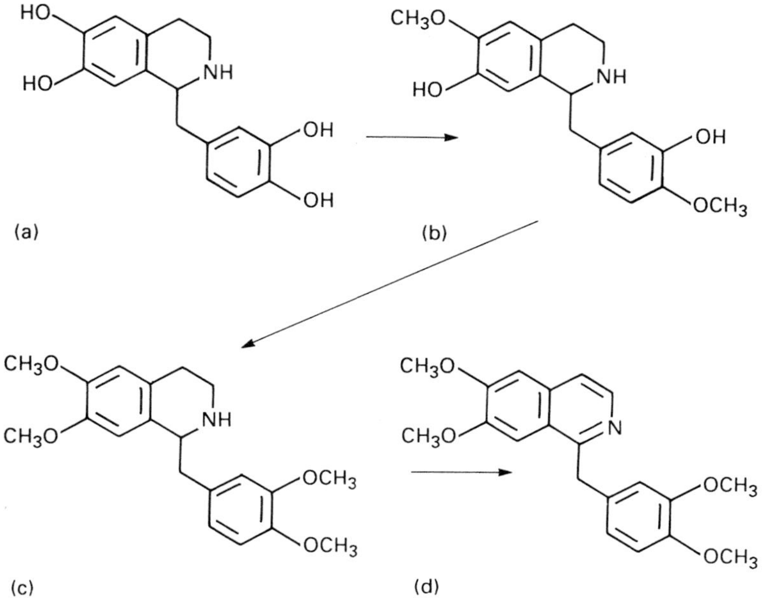

Fig. 7.46 (a) Norlaudanosoline, (b) norreticuline, (c) tetrahydropapaverine, (d) papaverine.

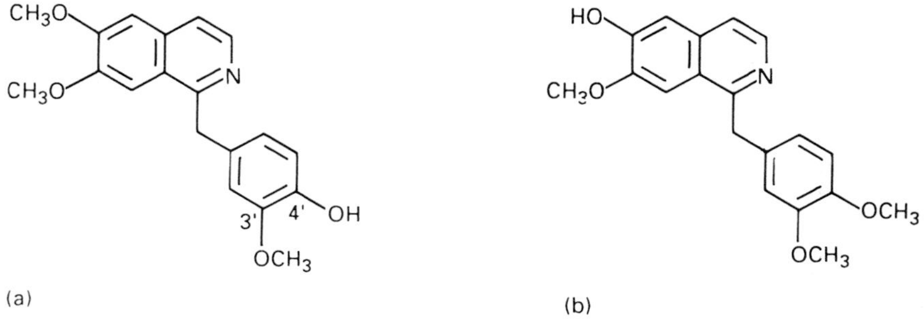

Fig. 7.47 (a) 4′-Monodemethylpapaverine, (b) 6-monodemethylpapaverine.

duced during the exponential phase of growth, the maximum concentration of 6-monodemethylpapaverine was reached when *S. alba* cells were in their stationary phase. However, these biotransformations which at first sight appear to be target-specific occur in different patterns with molecules similar to papaverine. For example isopapaverine, which is a structural analogue of papaverine, is preferentially demethylated at C-7 rather than C-6. In isopapaverine the C-7 has a more accessible position than in papaverine where the presence of the benzyl group at C-1 hinders C-7. These studies would therefore suggest that the *O*-demethylation

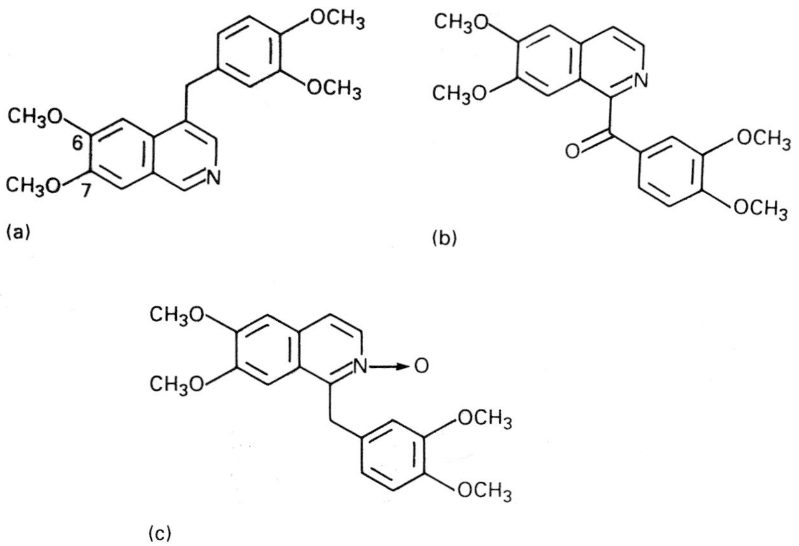

(a)

(b)

(c)

Fig. 7.48 (a) Isopapaverine, (b) papaveraldine, (c) papaverine *N*-oxide.

reactions are not specifically directed towards the benzylisoquinoline skeleton. The biotransformation of papaverine by *S. alba* cells is not confined to *O*-demethylation reactions. These cells also oxidize the benzyl group of papaverine to yield papaveraldine with a benzoyl group, as well as the nitrogen of the isoquinoline ring to give the *N*-oxide of papaverine (Fig. 7.48).

Another area where the application of biotransformation can help elucidate individual steps in biosynthetic pathways is that of tropane alkaloids. This group of alkaloids originate from the amino acid ornithine and occur principally in the species of the Solanaceae family. Medicinally active members of tropane alkaloids include (−)-hyoscyamine and hyoscine or scopolamine (Fig. 7.49). The drug atropine is a racemic mixture of (+) and (−)-hyoscyamine and is used to dilate the pupil of the eye in ophthalmic practice. In the whole plant, hyoscyamine is converted to hyoscine through the possible intermediates of 6,7-dehydrohyoscyamine and 6-β-hydroxyhyoscyamine. The epoxide ring of hyoscine is formed when the β hydrogens at positions C-6 and C-7 of the tropine ring are eliminated. Both hyoscyamine and hyoscine have been reported from cultured cells of the Solanaceae species albeit at significantly reduced yields than in the whole plant. One of the species reported to produce these tropane alkaloids in culture is *Anisodus acutangulus*. These suspension cultures are also able to convert added hyoscyamine to hyoscine in a two-step biotransformation process involving 6-β-hydroxyhyoscyamine as an intermediate. The work, carried out by scientists at the Chinese Academy of Medical Sciences, clearly showed that the decrease in the level of exogenous hyoscyamine was associated with a simultaneous rise in the

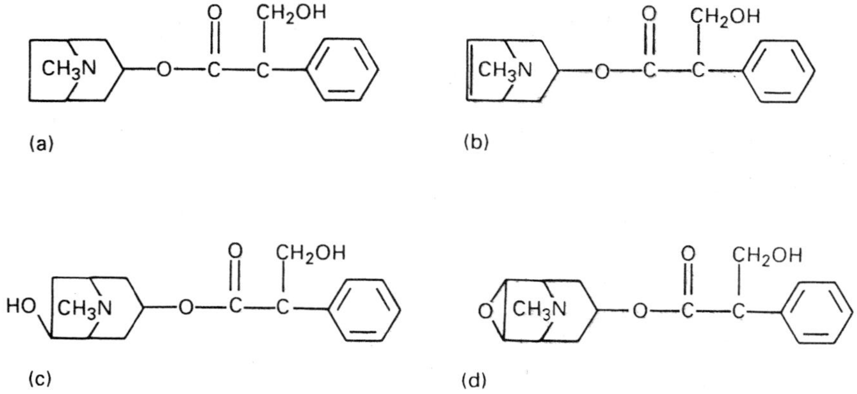

Fig. 7.49 (a) Hyoscyamine, (b) 6,7-dehydrohyoscyamine, (c) 6-hydroxyhyoscyamine, (d) hyoscine (scopolamine).

concentration of 6-β-hydroxyhyoscyamine. Although this conversion was evident in the early days of culture cycle, the formation of 6-β-hydroxyhyoscyamine gathered momentum when the cells reached their stationary phase and their dry weight started to decrease. Interestingly it was at this stage of the growth cycle, with the dry weight declining, that the second phase of the biotransformation, namely the conversion of 6-β-hydroxyhyoscyamine to hyoscine, took place. As in the first phase, the appearance and subsequent gradual rise in the level of hyoscine was accompanied by a drop in the concentration of 6-β-hydroxyhyoscyamine. Although *A. acutangulus* suspensions were thus shown to effect the two-phase biotransformation of hyoscyamine, the efficiencies of these stages were not equal. As the concentration of hyoscyamine was raised 10-fold, there was a fivefold increase in the amount of 6-β-hydroxyhyoscyamine in cultures, with the rise in hyoscine level not exceeding twofold. Similar results obtained with suspension cultures of *Hyoscyamus niger* cells suggest that the low levels of hyoscine produced from hyoscyamine via biotransformation are due primarily to low levels or reduced activity of the enzyme, or enzymes, responsible for the conversion of 6-hydroxyhyoscyamine to hyoscine. As a medicinal agent, hyoscine is preferred to hyoscyamine because of the latter's stimulatory action on the CNS. Consequently if the aim is to produce a desired product (e.g. hyoscine) from a precursor (e.g. hyoscyamine), then factors such as low enzyme levels constitute a disadvantage. However, in biphasic biotransformation such as the one described above, the insufficiency of the second stage can be an advantage if the intermediate substance (e.g. 6-hydroxyhyoscyamine) which will accumulate, is also a desirable product. In this particular example 6-hydroxyhyoscyamine fits the model because as an antispasmodic and analgesic agent it is extensively used in China and it has also been shown to improve microcirculation.

Immobilized Systems

During the past decade an increasing degree of interest has been focused on the immobilization of cultured plant cells and plant enzymes for the purposes of single-step biotransformations. For an account of plant cell immobilization objectives and procedures, see Chapter 8.

It has been argued by a number of workers that fixing plant cells on a support matrix represents a stage intermediate between the less differentiated system (suspension cultures) and the highly organized tissue structure of the whole plant. Thus a population of slow-growing cells with the possibility of partial differentiation induced by cell-to-cell contact would not only uncouple growth and secondary production – a condition favouring the latter in many systems – but would also extend the life span of the biocatalyst. A major disadvantage in biotransformation processes using immobilized plant cells is likely to be the permeability of entrapped cells to substrate and product. One solution to this formidable problem is to permeabilize cells, an approach which has gathered momentum in recent years. An alternative way of circumventing the cell membrane barrier and eliminating the possibility of unwanted side-reactions and product storage within the cell is to immobilize the enzyme responsible for a single biotransformation reaction. Alternative approaches to biotransformation processes using immobilized systems are schematically illustrated in Fig. 7.50.

Soon after the discovery of the potential in *Digitalis lanata* cell cultures to hydroxylate cardiac glycosides, these cells were entrapped in alginate gels in order to assess the efficiency of the process in an immobilized system. Although the biotransformation profile of digitoxin and β-methyldigitoxin was the same as in both suspension cultured and immobilized cells, the hydroxylating activity was reduced by a half in the latter. This apparent setback was however compensated by the fact that the immobilized cells displayed constant biotransformation activity for over 60 days, a considerable improvement on batch culture. The enzyme catalysing the biotransformation of cardiac glycosides is a 12β-hydroxylase located in the microsomes of *D. lanata* cells. The properties of the enzyme suggest that it is a cytochrome P-450-dependent monooxygenase requiring $NADPH_2$ as a cofactor. 12β-Hydroxylase has been immobilized in calcium alginate and has been shown to catalyse the product formation for more than 20 h, albeit with reduced hydroxylating rate compared to cell cultures.

In contrast to *D. lanata* cells the biotransformation efficiency of (−)-menthone to (+)-neomenthol (see Fig. 7.12) and of (+)-pulegone to (+)-isomenthone by immobilized *Mentha* cells was as high as that of freely suspended cells. In addition, not only were fewer monoterpene compounds (substrates and products) retained in the cells but certain secondary bioconversions observed in cell cultures, such as glycosylation, were substantially repressed in immobilized cells. Such encouraging results suggest that the problems of cell permeability and undesirable side reactions in plant cell immobilization may not be universal. A significant improvement in biotransformation capability of immobilized cells can be achieved in some systems by agents that cause division arrest in cells. For example, when *Mentha* cells were irradiated with γ-rays from a cobalt source they fully retained

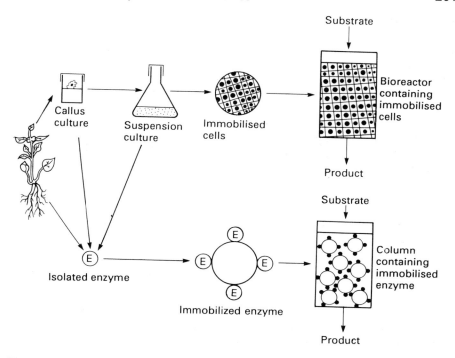

Fig. 7.50 Procedures for immobilization of plant cells and plant-derived enzymes.

their biotransformation potential, but more importantly they could be subjected to consecutive cycles of menthone to neomenthone conversions thus extending the time-course of the process. The suppression of mitotic activity without causing cell death in immobilized systems can have two distinct advantages:

- It should eliminate unnecessary cellular reactions and channel the saved metabolic energy towards transformation activity
- It should reduce the pressure inside the immobilization matrix caused by cell proliferation.

Amongst alkaloids, those belonging to the benzylisoquinoline family have been the subject of a number of immobilization studies. The Japanese scientist T. Furuya and his group who had successfully demonstrated the bioconversion of codeinone to codeine (see Fig. 7.43) by cell suspension cultures of, and a cell-free system from, *Papaver somniferum*, also assessed their biotransformation activity when immobilized in calcium alginate beads (Furuya *et al.*, 1984). In shake-flasks the immobilized cells retained their biological activity for over six months and when placed in a column bioreactor remained functional for one month. More importantly, about 90% of the product (codeine) was recovered in the bathing medium. The conversion of thebaine to codeine via codeinone has been proposed

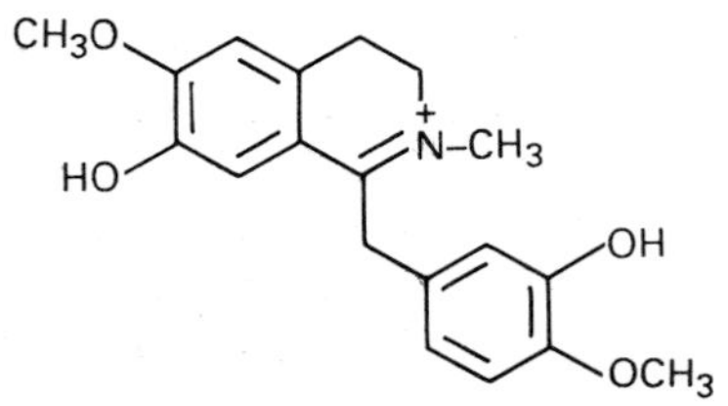

Fig. 7.51   Structure of 1,2-dehydroreticuline.

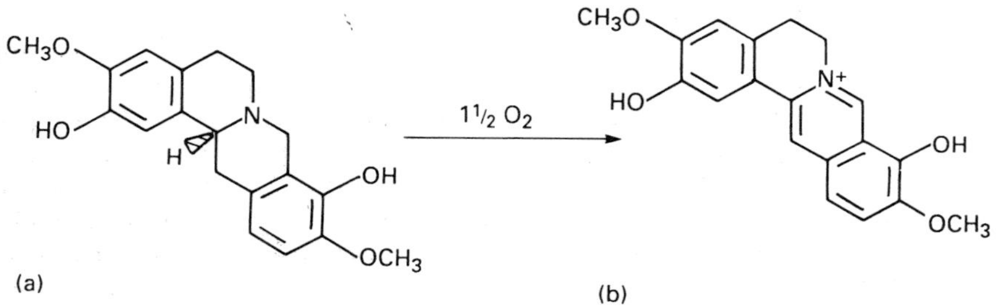

Fig. 7.52   Conversion of (a) (*S*)-tetrahydroprotoberberine [(−)-(*S*)-scoulerine] to (b) protoberberine by (*S*)-tetrahydroprotoberberine oxidase (STOX).

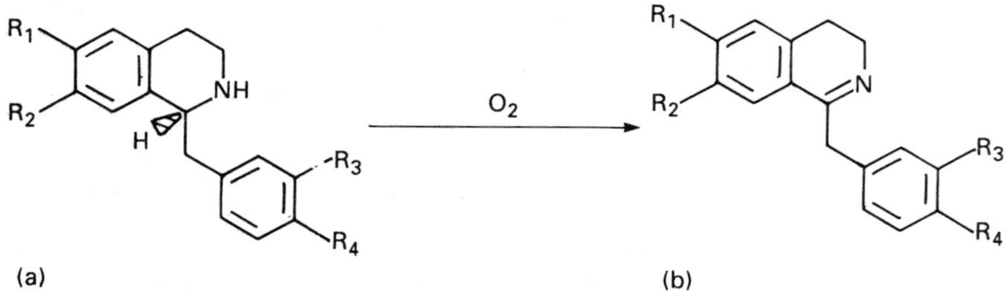

Fig. 7.53   Conversion of (a) 1-(*S*)-benzylisoquinoline to (b) 1,2-dehydroisoquinoline protoberberine by (*S*)-tetrahydroprotoberberine oxidase (STOX).

as a valuable process in the biotechnological production of plant metabolites, and successful immobilization of *P. somniferum* has brought the realization of that process one step nearer.

A crucial step in a future biotechnological synthesis of morphinan alkaloids is the biotransformation of (*S*)-benzylisoquinolines into their corresponding dehydro forms. Dehydroreticuline (Fig. 7.51) has been shown to be an intermediate in the interconversion of (*S*) and (*R*) forms of reticuline (see Fig. 7.41). The enzyme (*S*)-tetrahydroprotoberberine oxidase or STOX which converts (*S*)-tetrahydroprotoberberine to protoberberine (Fig. 7.52) can also oxidize (*S*)-benzylisoquinolines to their 1,2-dehydro forms (Fig. 7.53). STOX was first isolated and characterized by M.H. Zenk and his group in Munich. They have recently (Amman and Zenk, 1987) obtained the enzyme from cell suspension cultures of *Berberis wilsoniae* and immobilized it by several different methods. When compared with immobilized *B. wilsoniae* cells and the soluble enzyme, the overall performance of immobilized STOX was better with its stability enhanced 50-fold over that of soluble enzyme.

The immobilization of plant cells and selected plant enzymes has now become an established technique and increasing number of these biocatalysts have been used in biotransformation studies. Future research will surely be directed towards the quest for new approaches and materials as well as the further development and optimization of existing techniques. As far as biotransformation by immobilized systems is concerned, an important area of this research effort should be the scale-up of these systems and development of bioprocesses for the production of specialized chemicals.

References and Further Reading

Amman, M. and Zenk, M.H. (1987). Preparation of dehydrobenzylisoquinolines by immobilised ((*S*)-tetrahydro-protoberberine oxidase from plant cell cultures. *Phytochemistry* **12**, 3235–3240.

Berlin, J. (1986). Secondary products from plant cell cultures. In *Biotechnology* Vol. 4 (H. Pape and H-J Rehm, eds) VCH, Weinheim, pp. 629–658.

Braemer, R., Tsoutsias, Y., Huabielle, M. and Paris, M. (1987). Biotransformation of quercitin and apigenin by a plant cell suspension culture of *Cannabis sativa. Planta Medica* **53**, 225–226.

Bridel, M. and Lavieille, R. (1931). Sur le principe sucré des feuilles de Kaá-hê-è (*Stevia rebaudiana* Bertoni). *Comptes Rendues Hebdomedaires des Séances de l'Académie des Sciences, Paris* **192**, 1123–1125.

Brodelius, P. (1983). Production of biochemicals with immobilised plant cells: possibilities and problems. *Annals of the New York Academy of Sciences* **413**, 383–393.

Furuya, T., Nakano, M., and Yoshikawa, T. (1978). Biotransformation of (*RS*)-reticuline and morphinan alkaloids by cell cultures of *Papaver somniferum. Phytochemistry* **17**, 891–893.

Furuya, T., Yoshikawa, T., and Taira, M. (1984). Biotransformation of codeinone to codeine by immobilised cells of *Papaver somniferum. Phytochemistry* **23**, 99–1001.

Hosel, W. (1981). Glycosylation and glycosidases. In *Biochemistry of Plants*, Vol. 7, Academic Press, New York, pp. 725–753.

Kurz, W.G.W. and Constabel, F. (1979). Plant cell cultures, a potential source of pharmaceuticals. *Advances in Applied Microbiology* **25**, 209–240.

Mizukami, H., Terao, T., Miura, H., and Ohashi, H. (1983). Glucosylation of salicyl alcohol in cultured plant cells. *Phytochemistry* **22**, 679–680.

Reinhard, E. and Alfermann, A.W. (1980). Biotransformation by plant cell cultures. *Advances in Biochemical Engineering* **16**, 49–83.

Rosevear, A. and Lambe, C.A. (1985). Immobilised plant cells. *Advances in Biochemical Engineering* **16**, 49–83.

Sandermann, H., Jr., Diesperger, H., and Scheel, D. (1977). Metabolism of xenobiotics by plant cell cultures. In *Plant Tissue Culture and its Biotechnological Application* (W. Barz, E. Reinhard, and M.H. Zenk, eds), Springer-Verlag, Berlin, pp. 178–196.

Steck, W. and Constabel, F. (1974). Biotransformations in plant cell cultures. *Lloydia* **37**, 185–191.

Stohs, S.J. and Staba, E.J. (1964). Production of cardiac glycosides by plant tissue culture. IV. Biotransformation of digitoxigenin and related substances. *Journal of Pharmaceutical Science* **54**, 56–64.

Chapter 8

The Immobilization of Plant Cells

ALAN SCRAGG

Introduction

Plant cell culture has for some time been considered as an alternative method for the production of phytochemicals to their extraction from plantation-grown crops. Characteristics of plant cell cultures, such as slow growth, mean that, at least initially, the compounds produced should be of high value ($500–1000 kg^{-1}) and low volume. One of the limiting factors in the development of a commercial production system using plant cell culture has been production costs. The use of high biomass levels for extended periods would be one method of increasing productivity and hence reducing the costs. This can be achieved by the immobilization of plant cells.

The immobilization of cells and enzymes has received increasing attention, and has been used to produce amino acids and high-fructose syrups. The immobilization of microbial cells is not new; it has been known for some time that cells will adhere to many surfaces in nature. The development of film culture of micro-organisms has been used in the traditional production of vinegar, where *Acetobacter* cells are immobilized on oak chips, and in the trickle-beds used in waste water treatment.

Immobilization has been defined as a technique which confines a catalytically active enzyme or cell within a reactor system and prevents its entry into the mobile phase which carries the substrate and product. The first reported immobilization of plant cells was by Brodelius and colleagues in 1979 who entrapped *Catharanthus roseus* and *Daucus carota* cells in alginate beads. Since then immobilized plant cells have been used for a wide range of reactions which can be divided into three groups; bioconversions or biotransformations, synthesis from precursors, and the *de novo* synthesis of compounds (Fig. 8.1).

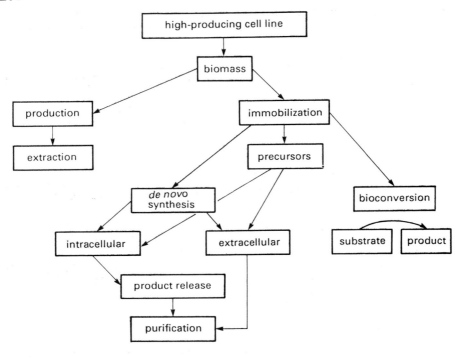

Fig. 8.1   Use of immobilized plant cells.

Advantages of Plant Cell Immobilization

The benefits of immobilization vary in importance, depending on whether the biocatalyst is a single enzyme, a micro-organism, or a large plant or animal cell, but can be represented by six criteria (Table 8.1).

RETENTION OF BIOMASS OR CATALYST

The retention of biomass enables its continuous reutilization as a production system, a definite advantage with slow-growing plant cells. The maintenance of a stable, active population of slowly- or non-dividing cells is the key to success. Cells of *Papaver somniferum* have remained stable and active for up to six months.

HIGH BIOMASS LEVELS

The immobilization of cells allows the use of a higher biomass level than would be possible with a free cell suspension, because of the limitations of mass transfer and settling. Bead densities of 110 g dry weight/L have been obtained with alginate-entrapped cells, in contrast to values of 30 g dry weight/L for freely suspended cells.

Table 8.1 Advantages of plant cell immobilization

- Retention of biomass
- High biomass levels
- Separation of cells from medium
- Allows a continuous process
- Uncoupling of growth and product formation
- Reduces problems such as aggregates, wall growth, or foaming

The high cell density allows a reduction in contact time in a packed bed catalyst leading to an increased volumetric productivity and a more compact system.

SEPARATION OF CELLS FROM PRODUCT

Immobilization separates the cells from the medium, and, if the product is extracellular, it can simplify downstream processing. The retention of the biomass removes the separation step and allows a continuous removal of product, making the recovery of labile or inhibitory products possible.

CONTINUOUS NATURE OF THE PROCESS

Immobilization allows a continuous process to be developed which increases the volumetric productivity and allows the removal of metabolic inhibitors.

SEPARATION OF GROWTH AND PRODUCT FORMATION

Immobilization allows the uncoupling of growth and production which permits product optimization without affecting growth. This can be achieved by using certain phytohormones which encourage secondary product formation but suppress growth, a condition not easily possible with freely suspended cells. In such a manner cell growth can be controlled and product formation encouraged.

PROBLEMS OF GROWTH

Immobilization can reduce some of the physical problems associated with the cultivation of plant cells. The formation of 'meringue' wall growth, the formation of aggregates, and susceptibility to mechanical damage are problems which do not affect immobilized systems.

Methods of Immobilization

Three main types of immobilization have been used for plant cells: adsorption, covalent attachment, and entrapment. Plant cell suspensions contain mainly cell

Table 8.2   Methods of immobilization

• Adsorption		
• Covalent attachment		
• Entrapment		
(a) Natural polymers		Alginate
		Agar
		Agarose
		k-Carrageenan
		Chitosan
(b) Synthetic polymers		Polyacrylamide
(c) Porous structure		Polyurethane foam
(d) Membranes		Hollow-fibre
		Flatplate

aggregates which can be up to 1 mm in diameter, so entrapment methods have been found to be most appropriate (Table 8.2).

ADSORPTION

Adsorption on to a solid support offers a simple and inexpensive method of immobilizing plant cells. Plant cells have been shown to bind to supports such as gelatin, agar, alginate, polypropylene, polystyrene, and glass. Biofilm culture offers the advantage that the cells are in direct contact with the growth medium, thus reducing nutrient limitations. In one example, *C. roseus* plant cells were adsorbed to calcium alginate in a packed-column reactor. Cytodex microcarrier beads coated with concanavalin A have been used to immobilize *Datura innoxia* protoplasts, and a further sophistication was to add a flexible link to the concanavalin A.

COVALENT COUPLING

Coupling agents such as glutaraldehyde and carbodiimides have been used to bind microorganisms to substrates such as glass. These agents have proved too reactive for plant cells, and only one example has been reported. Cells of *Solanum aviculare* were bound to polyphenylene oxide beads using glutaraldehyde for the production of glycoalkaloids.

ENTRAPMENT

Entrapment has been the most widely used method for immobilizing plant cells. A number of methods of entrapment have been used; formation of polymers, use of porous structures, and confinement behind membranes (see Table 8.2). For the entrapment of cells, natural polymers such as alginate, agar, and carrageenan, or synthetic polymers can be used. Table 8.3 shows that plant cells can be trapped in both natural and synthetic polymers. Alginate has clearly been the most popular

Table 8.3   The immobilization of plant cells using polymers

Natural polymers	Plant species
Alginate	Cannabis sativa
	Catharanthus roseus
	Coffea arabica
	Daucus carota
	Digitalis lanata
	Glycine max
	Lavandula vera
	Morinda citrifolia
	Mucuna pruriens
	Nicotiana tabacum
	Papaver somniferum
	Silybum marianum
	Ipomoea sp.
	Thalictrum rugosum
	Tagetes minuta
Agarose	Catharanthus roseus
	Daucus carota
	Datura innoxia
	Glycine max
	Thalictrum rugosum
Agar	Catharanthus roseus
	Daucus carota
	Glycine max
	Tagetes minuta
k-Carrageenan	Catharanthus roseus
	Daucus carota
	Tagetes minuta
Chitosan	Amaranthus tricolor
	Asclepias syriaca
	Apium graveolens
Synthetic polymers Polyacrylamide	Capsicum frutescens

because of its mild polymerization conditions. The more drastic conditions required for polyacrylamide have restricted the use of this polymer (Table 8.4). The effects of various methods of immobilization on the viability of plant cells are shown in Table 8.5. Alginate is a polysaccharide consisting of guluronic and mannuronic acids, and polymerization occurs when calcium ions form a bridge between guluronic acid units. It has the advantage that the gel can be formed at

Table 8.4 Polymers used in plant cell immobilization

Polymer	Polymer concentration (%)	Methods of polymerization
Alginate	2.5	Ionic cross-linking (0.1M Ca^{2+})
Carrageenan	1.5	Cooling from 50 to 20°C and ionic cross-linking with potassium
Agar	2.0	Cooling from 50 to 20°C
Agarose	2.5	Cooling from 40 to 20°C
Alginate and gelatin	2.0	Ionic cross-linking (0.1M Ca^{2+}) and chemical cross-linking (1% glutaraldehyde)
Agarose and gelatin	2.0	Cooling from 40 to 20°C and chemical cross-linking (1% glutaraldehyde)
Polyacrylamide	15.0	Polymerization with cross-linker

Table 8.5 Comparison of the effects of various immobilization methods on *Catharanthus roseus* cells. (Reproduced from Brodelius, 1980, with permission of the publisher)

Method of immobilization	Plasmolysis	Respiration	Cell growth
Alginate	+	+	+
Carrageenan	+	+	+
Agar	+	+	+
Agarose	+	+	+
Gelatin	+	−	−
Alginate and gelatin	+	−	−
Agarose and gelatin	+	−	−
Polyacrylamide	−	−	−

+ indicates a positive response
− indicates that the cells did not undergo plasmolysis, respiration or growth

room temperature and dissociates upon addition of phosphate ions. Figure 8.2 shows a simple method for the production of alginate beads containing plant cells.

Polyurethane foam has been used to immobilize a range of cell lines, including *Capsicum frutescens*, *Daucus carota*, *Cinchona pubescens*, *Beta vulgaris*, *Humulus lupulus*, and *C. roseus*. Nylon pan scrubbers have also been used to immobilize *Beta vulgaris* and *Humulus lupulus*. The cells are immobilized in these matrices either by flowing cells and medium through the foam or by adding the sterile foam to a growing

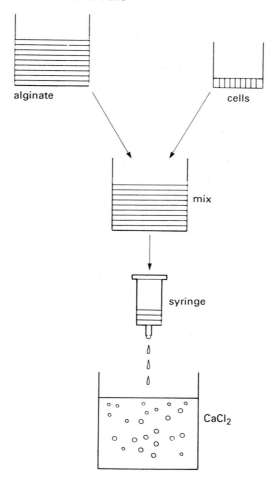

Fig. 8.2 The production of calcium alginate beads containing plant cells.

culture. The cells are obviously trapped due to the aggregate sizes, but the cells also become attached to the porous structures via a polysaccharide-based mucilage.

CONFINEMENT

The confinement of plant cells behind semi-permeable membranes such as those in hollow-fibre units has been used to separate the cells from the growth medium. The hollow-fibre units can be sterilized and the concentrated plant cells can be introduced into the shell side of the units (Fig. 8.3). The medium can be circulated through the fibre lumen and aerated using a separate reservoir. Flat-bed membrane reactors have also been used to immobilize plant cells.

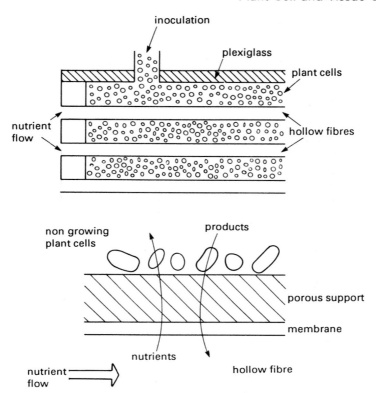

Fig. 8.3   Plant cell suspension entrapped in a hollow-fibre cartridge.

Properties of Immobilized Cells

VIABILITY

The effects of immobilization techniques on cell viability are shown in Table 8.5, in which respiration has been used as an indication of cell viability. However, as the cell density increases in the beads, the relative level of respiration decreases (Fig. 8.4). Thus the support material, cell loading, and bead diameter can all influence the respiration of entrapped cells. Oxygen concentration profiles within a bead show oxygen limitation which can slow growth and decrease oxygen requirement, but the effect of such limitation on secondary product formation is unknown.

GROWTH

Many of the methods used to estimate growth, such as increase in dry weight, or fluorescein diacetate staining, cannot easily be carried out on immobilized plant

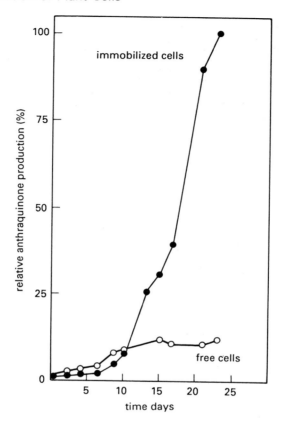

Fig. 8.4 Anthraquinone production by freely suspended and immobilized *Morinda citrifolia* cells (Reproduced from Brodelius and Nilsson, 1983, with permission of the publishers).

cells except perhaps with alginate which can be easily dissolved. A non-destructive method of measuring free space and cell volume in immobilized systems has been developed using tritiated water and radioactive mannitol. The tritiated water penetrates the cells whereas the mannitol is excluded giving a measure of intact cell volume and free space. Phosphorus NMR has also been used to follow the intracellular metabolism of immobilized cells.

Although the immobilized cells need to retain viability, the continued growth of the cells can be a limitation in the use of immobilization of plant cells because continued growth of cells can rupture structures such as alginate beads or hollow fibre units. Therefore, attempts are being made to formulate media which will maintain viability without growth over an extended period.

Table 8.6 Bioconversions by plant cells

Plant species	Immobilization	Substrate	Product
Digitalis lanata	Alginate	Digitoxin	Digoxin
	Alginate	Methyldigitoxin	Methyldigoxin
Daucus carota	Alginate	Digitoxigenin	Periplogenin
Catharanthus roseus	Agarose	Cathenamine	Ajmalicine
Mentha sp.	Polyacrylamide	(−)Menthone	(+)Neomenthol
Mucuna pruriens	Alginate	Tyrosine	DOPA
Papaver somniferum	Alginate	Codeinone	Codeine

BIOTRANSFORMATION

The use of immobilized plant cells for biotransformations (Chapter 7) has covered a wide range of reactions including hydroxylations, methylations and glycosylations (Table 8.6). The best-studied system has been the biotransformation of cardiac glycosides by *Digitalis lanata* and *Daucus carota*. The efficiency of conversion of digitoxin to digoxin by *D. lanata* cells was the same as freely suspended cells and was maintained for 33 days.

PRODUCT FORMATION FROM PRECURSORS

The activity of immobilized plant cells in the conversion of added precursors can be seen in Table 8.7, and at present this has only been used on a limited scale.

DE NOVO PRODUCT FORMATION

Immobilized plant cells have been shown to retain their ability to produce secondary products, and in some cases an increase in productivity relative to freely suspended cells has been seen (Table 8.8). In the case of *Morinda citrifolia*, cells in alginate beads produced 10-fold more anthraquinones than the suspension cells (see Fig. 8.4). The production of secondary product has also been extended by immobilization for up to 40 days.

PRODUCT RELEASE AND RECOVERY

Central to the success of an immobilized cell system is the release of the products or uptake of precursors. However, many of the secondary products formed in plant cells are not released from the cells but are stored in the vacuole. Some cells do spontaneously release products, and immobilization has been reported to facilitate the release of some products. It has been found with *C. roseus* and *N. tabacum* that changing the medium regularly promotes release of certain products. Constant

Table 8.7 Conversion of precursors by immobilized cells

Plant	Polymer	Precursors	Product
Capsicum frutescens	Polyurethane foam	Isocaptic acid	Capsaicin
Catharanthus roseus	Alginate	Tryptamine and secologanin	Ajmalicine

Table 8.8 *De novo* synthesis of products by immobilized plant cells

Plant species	Entrapment	Product
Capsicum frutescens	Polyurethane	Capsaicin
Catharanthus roseus	Alginate	Ajmalicine
Coffea arabica	Alginate	Methylxanthines
Glycine max	Hollow-fibre	Phenolics
Lavandula vera	Alginate	Blue pigment
Morinda citrifolia	Alginate	Anthraquinones
Solanum aviculare	Polyphenylene oxide	Steroid glycosides
Amaranthus tricolor	Chitosan	Oxalate
Asclepias syriaca	Chitosan	Proteases
Thalictrum rugosum	Polyurethane	Berberine
Dioscorea deltoidea	Polyurethane	Diogenin

removal of the extracellular product has also been shown to promote product release, presumably by preventing feedback inhibition. A number of resins have been used to remove indole alkaloids and anthraquinones from the medium during growth, and of these the neutral polymeric resin XAD-7 was found to be most effective. A water-insoluble triglyceride has also been used to retain volatile terpenoids produced by *Thuja occidentalis*, and the same compounds have been removed using XAD-4 resin (Fig. 8.5). One problem with the continual removal of products is that these resins can often remove the auxins and cytokinins from the medium, resulting in poor growth of the cells.

Brodelius (1983) used a number of compounds or temperature changes to permeabilize *C. roseus* cells and thus facilitate product release. Of the various chemicals used including toluene and chloroform, dimethyl sulphoxide (DMSO) was found to be most effective, releasing some 85–90% of intracellular ajmalicine from *C. roseus* cells. Other methods to encourage release, such as changes in temperature, pH, or medium ionic strength have had limited success, although it has been suggested that transport of ajmalicine across the cell membranes of *C. roseus* cells can be achieved by pH changes or high ionic strength media.

Physical methods such as ultrasound and electroporation have also been employed for product release. Electroporation has been widely used to transfer

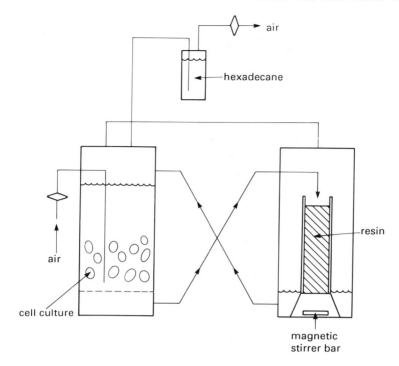

Fig. 8.5 The continuous removal of terpenoids from a *Thuja occidentalis* culture using a resin. The hexadecane was placed in the exit gas line to trap stripped-off volatile terpenoids. (Reproduced from Forche *et al.*, 1984, with permission of the publisher).

genes into plant protoplasts and takes advantage of the ability of electric pulses of an appropriate voltage to make the plasma membrane permeable (Chapter 3). Electroporation has been used with some success to release berberine from immobilized *T. regosum* cells.

Cell viability after treatment for permeabilization is also required to enable continuous operation with immobilized cells. A number of chemicals used for permeabilization have been shown to reduce viability drastically if used in the concentration required for full product release, and this response differs between cell lines.

Bioreactors

If immobilized plant cells are to be used for the production of various compounds or for biotransformations, careful consideration must be given to the choice of vessel or system in which they are used. Because of the complexities and range of

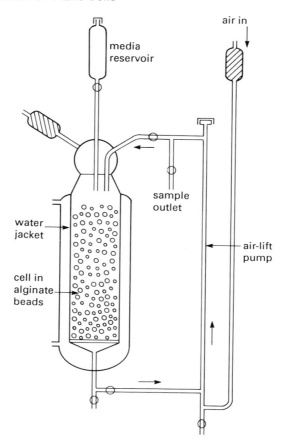

Fig. 8.6 A fluidized-bed reactor for immobilization of plant cells. (Reproduced from Velicky and Jones, 1981, with permission of the publishers.)

immobilization techniques, it is unlikely that one design will be suitable for all conditions. Unlike an immobilized enzyme system, immobilized whole plant cells must be provided not only with substrate but also with essential nutrients such as oxygen to maintain cell viability and biocatalytic activity.

The supply of oxygen, especially if the cells have a higher respiratory demand when immobilized, may cause problems such as foaming, circulation of beads, and inadequate sterility. Because of the long operation times due to slow product formation by plant cells, the reactor will probably be run at lower substrate concentrations than is usual for simple enzyme systems. This will mean that a simple flow-through would be unsuitable and a medium recycle should be adopted.

Immobilized plant cells can be formed into either particles or sheets. The

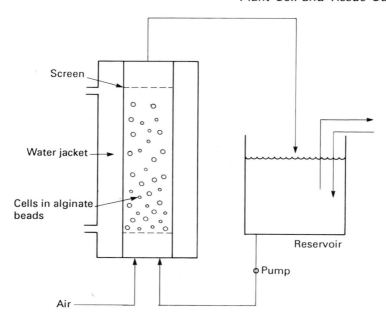

Fig. 8.7   A fluidized-bed reactor for plant cells with a recycle of medium.

particulate form is very versatile, allowing its use in simple systems such as shake flasks and more complicated systems such as packed-bed and fluidized-bed systems. Packed-bed reactors have been used for alginate beads containing cells of *D. carota* and *C. roseus*. In the first case aeration was by either direct bubbling into the column or by aerating a continuous supply of recycled medium (Fig. 8.6), whereas in the second system no specific aeration was provided, although provision for continuous extraction was provided (Fig. 8.7). A packed column using the strategy of medium recycle has been used for the production of steroid glycoalkaloids by *Solanum aviculare* cells. Fluidized-bed reactors have been used with *D. carota* cells in the conversion of sucrose to glucose and an air-lift bioreactor for alginate-immobilized tobacco cells.

Alternative bioreactor systems have been developed using cells immobilized in sheets which are wound into a coaxial configuration. A similar sheet immobilization, but in a flat-bed design, has been used for the production of phenolics from tobacco cells (Fig. 8.8). Hollow fibre membrane systems also represent another alternative system where the cells are entrapped. A hollow fibre membrane has been used to entrap *Glycine max* cells for phenolics production.

Conclusion

The immobilization of plant cells is now an established technique and a number of methods are available. At present the use of immobilized plant cells for secondary

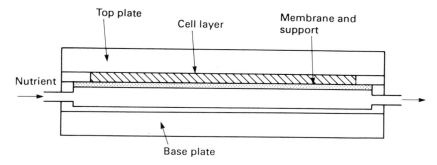

Fig. 8.8 A flat-bed reactor for the immobilization of plant cells.

product formation or biotransformation is at a very early stage in terms of media used, product formation, product release, and bioreactor design. However, results obtained to date suggest that the problems are not insuperable and that immobilization may give cost reductions for some processes.

References and Further Reading

Brodelius, P. and Nilsson, K. (1980). Entrapment of plant cells in different matrices. A comparative study. *FEBS Letters* **122**, 312–315.

Brodelius, P. and Nilsson, K. (1983). Permeabilization of immobilized plant cells resulting in release of intracellularly stored products with preserved cell viability. *European Journal of Applied Microbiology and Biotechnology* **17**, 275–280.

Brodelius, P., Deus, B., Hosboch, K. and Zenk, M.H. (1979). Immobilized plant cells for the production and transformation of natural products. *FEBS Letters* **103**, 93–97.

Forche, E., Schubert, W., Kohl, W., Hofle, G. (1984). Cell culture of *Thuja occidentalis* with continuous extraction of excreted terpenoids. Abstract Third European Congress on Biotechnology, Vol. 1. pp 189–192, Springer Verlag, Weinheim.

Knorr, D., Miazga, S.M., and Tentonico, R.A. (1985). Immobilization and permeabilization of cultured plant cells. *Food Technology*, October, 135–142.

Moo-Young, M. (ed.) (1988). *Bioreactor Immobilized Enzymes and Cells: Fundamentals and Applications.* Elsevier, London.

Rhodes, M.J.C. and Kirsop, B.H. (1982). Plant cell culture as sources of valuable secondary products. *Biologist* **29**, 134–140.

Velicky, I.A. and Jones, A. (1982). Bioconversion of gitoxigenin by immobilized plant cells in column bioreactor. Biotech Lett. **3**, 551–553.

Webb, C. and Mavituna, F. (1987). *Plant and Animal Cells: Process Possibilities.* Ellis Horwood, Chichester.

Webb, C., Black, G.M. and Atkinson, B. (1988). *Process Engineering Aspects of Immobilized Cell Systems.* Institute of Chemical Engineers, Rugby.

Yeoman, M.M. (1986). *Plant Cell Culture Technology.* Blackwell Scientific Publications, Oxford.

Chapter 9

Plant Cell Bioreactors

ALAN SCRAGG

Introduction

Mass cultivation of plant cell suspensions may be defined as the cultivation of plant cells in volumes above those normally used in shake flasks, which is generally about 1 L maximum. In practice, this means the culturing of plant cells in fermenters (also called bioreactors). There are two main advantages of cultivating plant cells in bioreactors.

First, the use of a bioreactor allows closer control and monitoring of cultural conditions than is possible using shake flasks. This is particularly true of the gaseous components such as oxygen or carbon dioxide which can be monitored either in dissolved form with probes, or as inlet or exit gas streams. In addition, a bioreactor offers a well-mixed system from which representative samples can be taken, and a system from which large (100 mL) samples can be removed for product analysis.

The second reason for the culturing of plant cells in bioreactors is to provide the information required for the scale-up of the culture in order to develop a commercial process. The mass cultivation of plant cells has been proposed as an alternative system to the normal plantations for the supply of high-value, low-volume phytochemicals such as the pharmaceuticals ajmalicine or codeine. Examples of possible, economically viable products are shown in Table 9.1. The first commercial process was announced by the Mitsui Petrochemical Company in 1983 where shikonin, a dye and pharmaceutical, was produced by cultures of *Lithospermum erythrorhizon*. Although cultural conditions for growth and product formation can be optimized in shake flasks, the response of some cell lines to growth in bioreactors cannot be predicted and upon scale-up these conditions cannot all be maintained at the same level. Upon scale-up some cultural parameters change

Table 9.1 Possible plant cell culture products suitable for industrialization

Compound	Use	Cost per kilogram ($)
Ajmalicine	Pharmaceutical, circulatory disease	1500
Codeine	Pharmaceutical, sedative	650
Digoxin	Pharmaceutical, heart stimulant	3000
Diosgenin	Pharmaceutical, steroid production	670
Jasmine	Fragrance	5000
Quinine	Food, bittering agent, antimalarial	100
Rose oil	Fragrance	3300
Saffron	Food, colouring and flavouring	10 000
Shikonin	Pharmaceutical, dye, antibacterial	4500
Spearmint	Flavour, fragrance	30

with the cube of the volume, whereas others increase only with the square of the volume. Thus before scale-up can be carried out the important parameters for growth and product formation in bioreactors must be determined.

The large-scale growth of plant cell suspensions started in 1959 with NASA-sponsored research on the possibility of using the cultures to supply food during space flight. The first vessels used were large carboys or bottles which were either rolled or bubbled to give good mixing. These makeshift bioreactors were soon replaced by stainless steel bioreactors which were used either as stirred-tank bioreactors, that is fitted with a motor and agitator, or as bubble tanks where sparged air was used to both aerate and mix the culture. These developments were mainly carried out in Japan, where there was a great deal of interest in the cultivation of tobacco cells for the production of low-tar, high-nicotine tobacco. The stirred-tank bioreactors were run at slow (50–100 r.p.m.) stirrer speeds to reduce shear stress, and culture volumes up to 6500 L were used. However, limited success was achieved with the growth of other plant cell cultures in stirred-tank bioreactors. The reason often given for this lack of success was that the high shear stress developed in the stirred-tank bioreactor broke up the plant cells, thus preventing growth.

In the early 1970s the adoption of an alternative bioreactor design, the air-lift bioreactor, for use in single-cell protein (SCP) processes, suggested their use for the cultivation of plant cells. The air-lift bioreactor is aerated and mixed using the inlet air stream, and the absence of a stirrer gives this type of vessel low shear characteristics. The air-lift bioreactor has been used to cultivate a number of cells lines (Table 9.2) and has proved very successful. However, as more information has become available concerning the cultivation of plant cells in bioreactors it has been found that plant cells are more robust than was expected; as a consequence it now may be possible to grow a wider range of plant cells in stirred-tank bioreactors than was first considered.

Table 9.2 The range of cell lines grown in bioreactors of different designs

Bioreactor system	Capacity (L)	Cell line cultured	Date
Sparged carboy	3–10	*Ginkgo, Lolium, Rosa, Mentha, Zea mays, Hyoscyamus niger*	1959–1975
Bubble column	1.8–1500	*Glycine max, Nicotiana tabacum*	1971–1975
Stirred-tank	2–15 500	Morning glory, *Nicotiana tabacum, Glycine max, Petroselinum, Morinda citrifolia, Spinacia oleracea, Phaseolus vulgaris, Cudriana tricuspidata, Catharanthus roseus, Helianthus annuus, Coleus blumei, Picrasma quassioides*	1971– present
Air-lift	7–100	*Morinda citrifolia, Catharanthus roseus, Theobroma, Cudriana tricuspidata, Berberis wilsoneae, Helianthus annuus, Cinchona ledgeriana*	1977– present
Rotating drum	4–100	*Catharanthus roseus*	1983

Bioreactor Characteristics

A bioreactor or fermenter is a vessel constructed of glass or steel in which organisms are cultivated. It has to fulfil a number of requirements:

- A sterile environment
- Access for environmental monitoring e.g. pH, temperature, dissolved oxygen
- Provision for sampling
- Provision for addition of antifoam, acid, base, fresh medium, etc.
- Supply of air
- Mixing
- Temperature control, addition or removal of heat.

These requirements can be provided in various ways. The foremost design of bioreactor has been the stirred tank since the early days in the development of

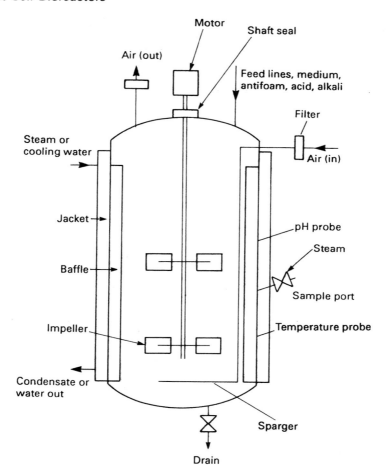

Fig. 9.1   A stirred-tank bioreactor.

penicillin production. The major features of the design are shown in Fig. 9.1 and the relative dimensions are shown in Fig. 9.2. Many of the cultures grown in bioreactors are aerobic, and the air required is normally blown or sparged in at the base of the bioreactor. Of the substrates supplied, oxygen is the least soluble. The impeller therefore has two functions: mixing of the culture, and break-up of the air bubbles to increase their surface area. As a result of the work of Rushton and co-workers (1969) a standard design of impeller, the turbine impeller, has been developed. The design of the turbine impeller and the flow pattern it achieves is shown in Fig. 9.3. More than one impeller can be used in a larger bioreactor.

The air-lift bioreactor was patented as early as 1955, but it gained prominence only during the investigation into SCP production in the late 1960s and early

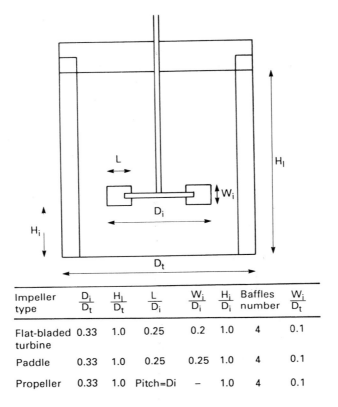

Impeller type	$\dfrac{D_i}{D_t}$	$\dfrac{H_l}{D_t}$	$\dfrac{L}{D_i}$	$\dfrac{W_i}{D_i}$	$\dfrac{H_i}{D_i}$	Baffles number	$\dfrac{W_i}{D_t}$
Flat-bladed turbine	0.33	1.0	0.25	0.2	1.0	4	0.1
Paddle	0.33	1.0	0.25	0.25	1.0	4	0.1
Propeller	0.33	1.0	Pitch=Di	–	1.0	4	0.1

Fig. 9.2   Relative dimensions of a stirred-tank bioreactor.

1970s. The air-lift bioreactor, as its name implies, is driven by air (Fig. 9.4). Air sparged into the base of the bioreactor lowers the density of the medium which rises up the draft tube pulling fresh medium in at the base, and therefore a flow is achieved. This type of bioreactor has no moving parts, no intrusion of impeller shafts with associated problems of seals, together with possible economic advantages as it does not require power for the input for the stirrer. A variety of designs of this type of bioreactor have been used to grow a wide range of micro-organisms.

However, it has become clear that plant cells in suspension are somewhat different from micro-organisms and therefore have different requirements. These are described in Table 9.3.

MIXING

Suspensions of both micro-organisms and plant cells require mixing in order to provide an even distribution of nutrients, a homogeneous suspension, and heat

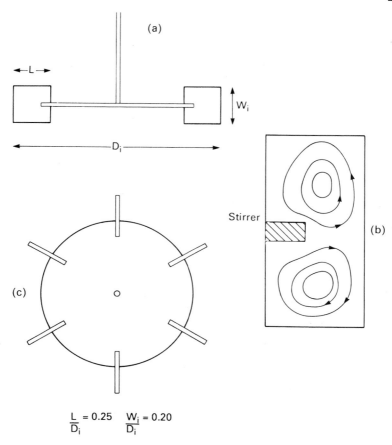

Fig. 9.3 (a) Relative dimensions of a turbine impeller; (b) pattern of flow in the reactor; (c) plan view of impeller.

and gas transfer. In a conventional bioreactor this is normally achieved by a motor-driven impeller. Early impeller systems were based on marine propellers or flat paddles, but the six-bladed flat disc turbine, the Rushton turbine, is now generally used (Fig. 9.3). Baffles are often used to break up the fluid flow further. Plant cells are large compared with micro-organisms (up to 100 μm in length) and generally occur as aggregates of up to 2 mm in diameter (Fig. 9.5). This large size causes a high settling rate in most plant cell suspensions, which will allow sediments to form in poorly mixed areas (or dead spaces) in the bioreactors. Therefore, good mixing is important with plant cell suspensions. However, plant cells are normally regarded as shear sensitive and the high shear stress developed by an impeller could cause break-up of cells or loss of cell viability. Thus when plant cells are grown in stirred-tank bioreactors a compromise is normally

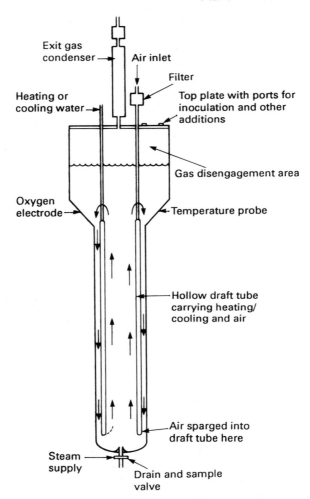

Fig. 9.4   An air-lift bioreactor.

achieved by removal of the baffles to stop cells being trapped on them, and a reduction in the stirrer speed to 50–150 r.p.m. Alternative impeller designs such as the marine propeller, anchor, or spiral have also been tried with some success (Fig. 9.6).

AERATION

Although plant cells are large, much of their volume consists of a vacuole so that the cytoplasmic content is relatively low. This low cytoplasmic content gives plant

Table 9.3 The characteristics of plant cell and microbial cultures

Characteristics	Microbial cell	Plant cell suspension	Consequences for plant cells
Size	2–10 µm	50–100 µm	Possibly shear sensitive; rapid sedimentation
Individual cells	Often	Not often, generally aggregated up to 1 mm in diameter	Rapid sedimentation possible micro-environment; low viscosity; large sample ports
Doubling time (growth rate)	1 h (rapid)	Days (slow)	Long culture runs, sterility maintenance
Inoculation density	Small	5–20%	Large inoculation vessels required
Shear stress	Insensitive	Sensitive	Slow stirred speeds required, may eliminate STR bioreactor
Aeration	High 1–2 vol/vol/min	Low 0.2–0.3 VVM	Low aeration required low $K_l a$ adequate
Variability	Can be stable	Can vary greatly	Culture characteristics not stable; may require continual selection
Product formation	Often into medium	Into vacuole	Cells need harvesting, no continual extraction from the medium

cells a low overall metabolic activity unit volume. This coupled with a slow growth rate makes the oxygen requirements of plant cells much lower at about 1–3 mmol O_2/g/h than the 10–100 mmol O_2/g/h for micro-organisms. Oxygen is rather insoluble in water (saturation value 7 p.p.m. at 25 °C) and therefore its supply is normally the limiting factor in aerobic micro-organism cultures. This does not appear to be true for plant cell cultures. Therefore, in plant cell suspensions, mixing may be more important than aeration. Air is normally sparged in at the base of the bioreactor to form a stream of bubbles which can be reduced in size by the impeller. The oxygen present has to dissolve in the medium in order to be transferred to the plant cells. The rate of oxygen transfer can be affected by a number of resistances, including gas film resistance, intrafacial resistance at gas–liquid interface, liquid film resistance between the interphase and the bulk liquid, liquid phase resistance for the transfer of oxygen from the bulk liquid to the film around the cell, liquid film resistance, intracellular or intra-aggregate resistance, and resistance due to oxygen consumption within the cell (Fig. 9.7). The overall resistance is the sum of these individual resistances and depends on the medium composition and density, and the activity of the cells. The major

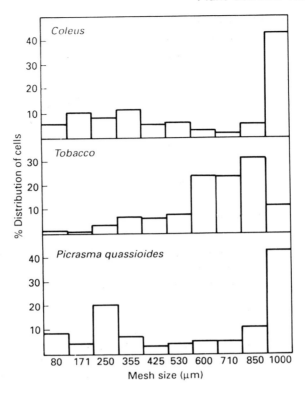

Fig. 9.5   Size distributions of plant cell aggregates.

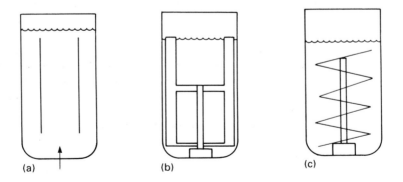

Fig. 9.6   Impeller designs for air-lift and stirred-tank (STR) reactors. (a) air-lift, 0.6 vol/vol/min (vvm) ; (b) anchor impeller in STR, 0.6 vvm and 40 r.p.m.; (c) spiral impeller in STR, 0.6 vvm and 50 r.p.m.

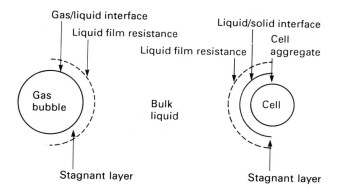

Fig. 9.7 Rate of oxygen transfer in a plant cell culture.

resistance is the liquid film around the bubbles and thus the overall oxygen transfer coefficient (K_L) is very similar to the liquid film transfer coefficient.

The rate of mass transfer or the volumetric oxygen uptake rate (OUR) can be defined as:

$$\text{OUR} = K1_a\,(C^* - C_L) \tag{1}$$

where C^* is the equilibrium concentration of dissolved oxygen (mmol), C_L is the oxygen concentration in liquid (mmol), K_l is the liquid side mass transfer coefficient (cm/s), a is the interfacial area, surface area of bubbles (cm^2), and OUR is the volumetric oxygen uptake rate.

The measurement of the mass transfer coefficient ($K_l a$), the combination of the mass transfer coefficient and interfacial area for oxygen, is a useful method of characterizing a bioreactor in terms of its ability to transfer oxygen to cultures. $K_l a$ can be determined by a number of methods, but one of the most popular is the dynamic gassing-in method. This is normally carried out in vessels without cultures and is based on measuring the transient change of dissolved oxygen when the bioreactor is aerated. The dissolved oxygen is removed from the vessel by sparging with an inert gas such as nitrogen (Fig. 9.8). To evaluate $K_l a$, log $(C^* - C_L)$ is plotted against the time which yields a straight line where $K_l a$ equals the negative of the slope. Low $K_l a$ values have been shown to reduce the growth of plant cells, whereas growth of other cultures have been shown to be inhibited at high $K_l a$ values. In general, plant cell suspensions are capable of growth at $K_l a$ values of around 10/h whereas micro-organisms require considerably more – around 100–1000/h. During growth of plant cell suspensions in bioreactor the levels of oxygen can drop to very low values (Fig. 9.9).

SHEAR SENSITIVITY AND RHEOLOGY

Because of their large size, extensive vacuole, and rigid cell wall, plant cells have been regarded as sensitive to shear stress. Recent studies of the shear sensitivities of

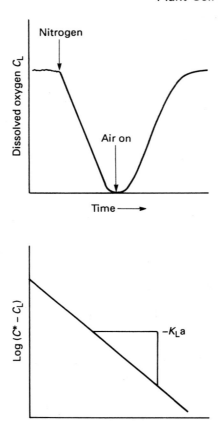

Fig. 9.8 Effect of sparging on rate of oxygen transfer.

some plant cell suspensions has shown that they are considerably more robust than was at first thought.

Suspensions of plant cells have been exposed for up to five hours in a small bioreactor (3 L) being agitated at 1000 r.p.m. which represents an average shear rate of 167/s. The small bioreactor was run under sterile condition so that samples could be removed at intervals, inoculated back into shake flasks, and growth followed. The ability to divide and grow has been used as a measure of the cells' viability. The normal measures are indirect; for example, the use of stains such as fluorescein diacetate and Evans blue only indicate an intact cell membrane. Of 12 plant cell cultures tested in the small bioreactor, only three proved sensitive to shear. This has meant that the cultivation of plant cell suspensions in stirred-tank bioreactors has been reconsidered. Recently cultures of *C. roseus* have been grown in a similar 3 L stirred-tank vessel at both 500 and 1000 r.p.m. with reduction in growth rate and biomass at the higher value. A possible explanation of the

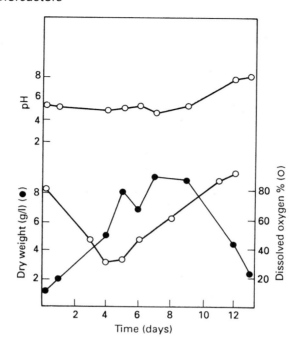

Fig. 9.9 Oxygen levels during the growth of a plant cell suspension.

difference between this work and the earlier results may be the improved cultures and culture techniques, and the use of higher shaker speeds (150 r.p.m.) for the stock shake-flask cultures such that selection for shear resistance has occurred.

At high concentrations (≥ 20 g dry weight/L) plant cell suspensions can appear to be viscous, and it has been suggested that at these biomass levels the viscosity would affect growth in air-lift bioreactors. The rheological behaviour of certain plant cell cultures has been determined; for example *Morinda citrifolia* showed shear thinning and thixotrophic behaviour whereas *Cudrania tricuspidata*, *Vinca rosea*, and *Agrostemma githago* were non-Newtonian and pseudoplastic.

In these studies the rheological properties of the cultures could be estimated using normal rotational viscometers such as the cup-and-bob system. However, plant cell suspensions are in general highly aggregated (see Fig. 9.5). As a consequence of this particulate nature, the normal methods of measuring viscosity, i.e. the cone-and-plate or cup-and-bob viscometers are usually inadequate, because the aggregates settle rapidly or become trapped in the narrow gaps in the viscometer. Alternative designs of viscometer using either anchor or turbine impellers that eliminate the problems of settling have been used but they do not give a direct measurement of viscosity because of the turbulent flow produced. Using both alternative types of viscometer, plant cell suspensions have been shown to be non-Newtonian and pseudoplastic having a low viscosity in the order of

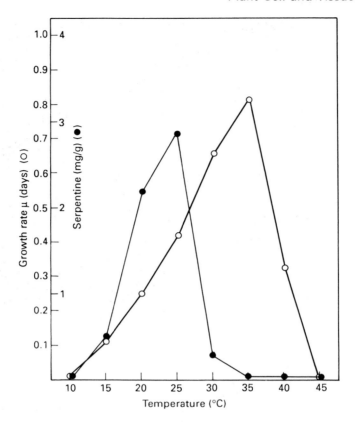

Fig. 9.10   Growth and product formation as affected by culture temperature for a suspension of *Catharanthus roseus* cells.

1–10 mPa/s. A similar situation existed when the viscosity of pelleted and mycelial fungal cultures were compared, and the pelleted cultures had a considerably lower viscosity than the mycelial culture. It would therefore appear that high viscosities in plant cell suspension may not be a major problem.

OTHER CULTURAL PARAMETERS

In general the pH of plant cell cultures is not controlled other than by adjusting the initial pH value to between 5 and 6 prior to inoculation. After this the pH is allowed to drift. The pH often drops soon after inoculation to values of 4–5 rising slowly to 5 to 6 or above as growth proceeds (see Fig. 9.9). The changes in pH are due to the preferential uptake of certain ions from the medium or an efflux of protons caused by this uptake. The effects of controlling pH are equivocal with

Fig. 9.11 'Meringue' formation during cultivation of tobacco cell suspensions in a bioreactor.

some evidence that it inhibits growth whereas other cultures require addition of alkali or acid addition to stop the pH reaching values which inhibit growth.

The temperature range for the cultivation of plant cells is normally between 25 and 30 °C. Little research has been carried out on the effect of temperature on growth and secondary product formation, but recently it has been shown that a *C. roseus* suspension culture has a growth optimum of 35 °C but a product (serpentine) formation optimum of 20 °C (Fig. 9.10).

Because of the slow growth of plant cells their heat evolution is low, so that in any bioreactor temperature control really involves the addition of heat to maintain 25 °C.

FOAMING AND WALL GROWTH

The formation of foam is quite common in microbial systems, where it is associated with protein in the medium. Foaming also occurs in some plant cell suspensions but its cause is perhaps not protein, as no proteins are added to the medium, but may be related to extracellular polysaccharide production. The formation of a stable foam in some cultures allows the build up of a thick crust or 'meringue' (Fig. 9.11). The meringue appears to consist of polysaccharides in which cells are trapped. The build-up of 'meringue' can be such that it captures a large percentage of the cultured cells and can stop the flow of culture in air-lift bioreactors.

Bioreactor Design for Plant Cell Suspensions

Few bioreactors have been designed specifically for plant cells but, with the differences and constraints associated with plant cell suspensions in mind, a number of microbial bioreactors have been modified.

STIRRED-TANK BIOREACTOR

Most existing laboratory or commercial bioreactors are of the stirred-tank design, and in general the following modifications have been carried out in order to grow plant cells in these bioreactors:

- The impeller speed is reduced to 50–150 r.p.m., and in some cases the turbine impeller replaced by a marine screw or paddle. Other designs of impeller have been used in an attempt to achieve low-shear mixing. These designs have been of the anchor or helical screw type of which the latter has proved very successful whereas the anchor impeller has a tendency to disrupt cell aggregates by trapping them between impeller and bioreactor wall (see Fig. 9.6)
- The removal of baffles, pH probes, or other probes not required reduces dead spots where the rapidly sedimenting cells can build up. The dissolved oxygen probe is often retained
- The sample ports are enlarged where required to about 1 cm to reduce blockages caused by the aggregated nature of plant cells, and dead spots where cells can settle are reduced as far as possible. For small laboratory bioreactors the sample ports are fitted with a pinch clip system or a system where air can be used to clear any blocked line (Fig. 9.12)
- A condenser is required to reduce water loss during the long culture runs (3–4 weeks). In most cases ports are available for the addition of sterile water, should this be required. With some cultures where foaming is a problem and the exit gas is monitored for either carbon dioxide or oxygen levels, a catch-pot is placed before the exit filter to stop carry over and blockage of this filter
- The existing temperature control can often be replaced if internal components are to be reduced to a minimum, by using a spiral half-round tube wrapped around the vessel and linked to a water bath and thermocirculator
- As can be seen from Fig. 9.9, pH is not normally controlled in plant cell cultures. In these cases the pH probe is normally removed and pH monitored from the daily samples. Recently we have grown plant cell suspensions which, without pH control, caused the medium pH to drop to levels low enough to kill the culture. Under these conditions pH control is required with a pH probe, a proportion controller and two pumps, one for acid and the other for alkali.

AIR-LIFT BIOREACTORS

Air-lift bioreactors have been successfully used to culture a wide range of plant cells (see Table 9.2). A number of designs can be found in the literature: home-

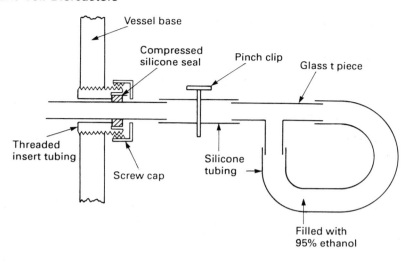

Fig. 9.12 Detail of a sampling port on a laboratory bioreactor.

made external loop, internal draft tube, all glass, or glass and metal. Commercial air-lift bioreactors are now available from a number of companies in a range of sizes from 2 to 100 L.

ALTERNATIVE BIOREACTOR DESIGNS

Recently alternative bioreactor designs to the stirred-tank bioreactor have been used to cultivate plant cell suspensions. These designs have been directed towards providing good mixing or aeration without imposing a high shear stress on the culture. The first was the rotating-drum bioreactor, where mixing is achieved by the rotating of the drum which has a number of baffles. This type of bioreactor has been used to grow cultures of *C. roseus* and *L. erythrorhizon* up to 1000 L in volume. A second bioreactor design has been based on Taylor–Couette flow where mixing was achieved by the vortices formed between two concentric rotating cylinders. The rotation speed was 200 r.p.m., and a 2.5 L bioreactor has been successfully used to grow *Beta vulgaris* cultures. The last design of bioreactor has used hydrophobic porous polypropylene hollow-fibre membranes to provide the oxygen required by the culture. Air passed through a coil of these membranes in the bioreactor allowed oxygen to diffuse into the culture, and mixing was by a slow speed magnetic stirrer. A 21 L version of the bioreactor has been used to cultivate *Thalictrum rugosum* to levels of 50 g/L.

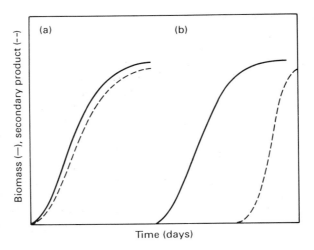

Fig. 9.13 Relation between growth and secondary-product formation.

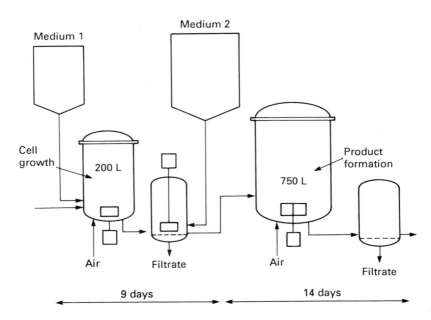

Fig. 9.14 The Mitsui shikonin process.

Plant Cell Processes

SINGLE-STAGE OR TWO-STAGE SYSTEMS

The mass cultivation of plant cells has been proposed as an alternative supply of plant material to plantation-grown plants. In a few cases such as tobacco and ginseng it is the cell mass that is required, but in the majority of cases a specific compound is required. These compounds are generally known as secondary products, compounds not essential for growth, and include the alkaloids, glycosides, and terpenes. The conditions under which these secondary products are formed influence the process and therefore how the bioreactor is used. Secondary product formation is sometimes linked to growth (Fig. 9.13), but in the majority of cases the product is formed after growth has ceased, or a change in culture medium or condition is required to induce product accumulation. Where the product is growth-associated, a simple batch process is sufficient. However, if a change in culture conditions or medium is required a two-stage process is required. The first stage allows rapid growth and the second product accumulation. Examples of a two-stage system are the production of serpentine in *C. roseus* demonstrated by Zenk, the production of shikonin by *Lithospermum erythrorhizon* (Fig. 9.14), and rosmarinic acid by *Coleus blumei* (Fig. 9.15). The *L. erythrorhizon* cells are grown rapidly in the first vessel, the medium is changed, and the red dye shikonin is synthesized in the second medium.

In the production of rosmarinic acid by *C. blumei*, the cells are grown rapidly to a high density by adding more sucrose at intervals during growth (fed-batch). The fed-batch culture avoids osmotic and feedback problems which may occur if very high initial sucrose levels are used. Once the cells have reached a high density, the medium is removed and replaced by sucrose alone which induces rosmarinic acid production (Fig. 9.15).

CONTINUOUS CULTURE

Continuous culture is an open system where the culture population is maintained in a continuous state of balanced growth by removing some of the culture and replacing it with fresh medium at the same rate. Continuous culture has been applied to plant cell cultures with some success, but because of the slow growth of plant cells, meringue formation, and maintenance of sterility it is not an easy system to use. However, it has been used at a working volume of 6500 L for the continuous production of tobacco biomass, but there are no reports of its use for the production of secondary metabolites.

Conclusion

The growth of plant cell suspensions in bioreactors still remains as a difficult technique but it is now possible to grow plant cells in bioreactors of $2->10,000$ volume and using both stirred-tank or air-lift designs. The major problems which

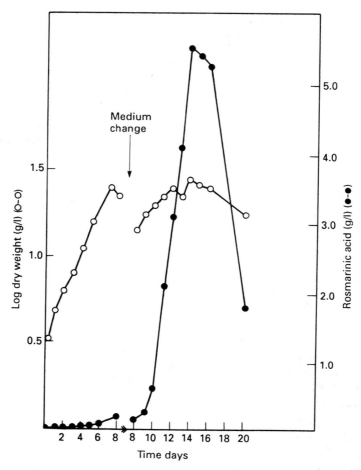

Fig. 9.15   Production of rosmarinic acid by *Coleus blumei*.

remain to be solved are the determination and maintenance of optimal conditions for product formation in bioreactors and the determination of what factors are important upon scale-up.

References and Further Reading

Anderson, C., Le Grys, G.A., and Solomons, G.L. (1982). Concepts in the design of large scale fermenters for viscous culture broths. *Biochemical Engineering* **Feb.**, 43–49.

Curtin, M.E. (1983). Harvesting profitable products from plant tissue culture. *Biotechnology* **1**, 649–657.

Dixon, R.A. (ed.) (1985). *Plant Cell Culture*. IRL Press, Oxford.

Morris, P. (1986). Regulation of product synthesis in cell cultures of *Catharanthus roseus*. IV: Effect of culture temperature. *Plant Cell Reports* **5**, 427–429.

Rushton, J.H., Costrick, E.W., and Everett, H.J. (1969). Power characteristics of mixing impellers. *Chemical Engineering Progress* **46**, 395–404.

Scragg, A.H. and Fowler, M.W. (1985). Mass culture of plant cells. In *Cell Culture and Somatic Cell Genetics of Plants* (I. Vasil, ed.) Academic Press, London, pp. 103–128.

Tanaka, H. (1981). Technological problems in cultivation of plant cells at high density. *Biotechnology and Bioengineering* **23**, 1203–1218.

Ulbrich, B., Wiesner, W., and Arens, H. (1985). Large-scale production of rosmarinic acid from plant cell cultures of *Coleus blumei* Benth. In *Primary and Secondary Metabolism of Plant Cell Cultures* (B. Deus-Neumann, W. Barz, and E. Reinhard, eds), Springer-Verlag, Berlin, pp. 293–303.

Wang, D.I.C., Cooney, C.L., Demain, A.L. *et al.* (eds) *Fermentation and Enzyme Technology*. Wiley, Chichester.

Glossary

Activity staining: Localization of enzymes on the basis of their catalytic properties, usually using a chromogenic substance.

Adventitious meristem: A meristem arising from disorganised cells.

Aglycone: The non-glycosylated analogue of a glycoconjugate.

Alkaloid: Typical alkaloids are basic, contain one or more nitrogen atoms, usually within a heterocyclic ring, and they usually have a marked physiological action on animals.

Anthicholinergic: Having the action of preventing the release of acetyl choline at the nerve endings of nerve fibres.

Anthocyanin: Glycoside of an anthocyanidin and responsible for the colours of blue, red and mauve flowers and fruits. Anthocyanidins are blue in alkaline solution and red at acid/neutral pH.

Anti-HIV: Anti human immunodeficiency (AIDS) virus.

Antimetabolite: A closely-related analogue of a metabolite (eg amino acid, nucleoside) that is not metabolized because of the structural differences. Such molecules usually competitively inhibit metabolism of the corresponding metabolite.

Anti-sense gene: Transcribed DNA having a complementary base sequence to the gene in question. The resulting RNA may inhibit translation of the original sense gene's mRNA by hybridization.

Antiserum: Blood serum from an animal inoculated with a given antigen. The serum is a crude preparation of the antibodies to the injected antigen.

Autonomous: Self-replicating.

Autotroph: If an organism does not require a particular essential growth substrate or cofactor to be added exogenously (i.e. it can synthesise it) it is said to be autotrophic with respect to that substrate. In the general sense, an autotroph can synthesise all its components from inorganic precursors.

Auxin: Plant hormone traditionally considered to promote cell enlargement and elongation eg. IAA (natural) and 2,4-D (synthetic).

Auxotroph: If an organism requires an exogenous supply of certain substance for growth (ie. it cannot synthesize it) then the organism is said to be an auxotroph with respect to that substance.

Axillary buds: Those formed in the axil (the angle between a leaf and the stem).

Axillary meristem: Meristem giving rise to axillary buds.

Back mixing: Back mixing means that any medium flow through or into a reactor is mixed with the bulk medium due to stirring or other forms of agitation.

Balanced growth: A term often applied to the growth of plant cells and which refers to that state of growth in which there is a parallel increase in both dry and wet weights of the culture.

Batch culture: The growth of cells in a fixed volume of medium. The cells undergo lag, logarithmic and stationary phases, the last occurring when the growth nutrients become depleted.

Bioreactor: A vessel used for a range of biological processes, usually the large-scale growth of cells.

Callus: A culture of cells on a solidifed growth medium.

Cardenolide: A type of cardio-active glycoside (C_{23}) containing a 5 membered lactone ring, e.g. digitoxigenin.

Carotenoid: A C_{40} terpenoid structure, biosynthesized in plant chloroplasts and derived from geranyl geranyl pyrophosphate. Carotenoids are often pigmented; yellow/red.

Cell division cycle: The sequence of events occurring during the division of a cell. The various phases include the synthetic (S) phase during which DNA is replicated and conspicuous pauses (G_1 and G_2).

Cell-free system: A cell homogenate containing the enzyme systems of the original cellular material.

Cell-line: A propagated cell culture having relatively stable characteristics. In extremely variable cultures, each individual flask may constitute a distinct cell line.

Cell vacuole: A subcellular organelle bounded by a membrane called the tonoplast. Vacuoles are present in all plant cells either as numerous small vesicles or fused together into larger units. Vacuoles have a range of functions including storage and degradation.

Chimeric gene: An artificially-constructed stretch of DNA containing components from separate genes. When the new gene is expressed, a hybrid protein can be produced. Alternatively the two original genes can be co-expressed.

Cofactor: A low molecular weight substance that is necessary for the normal functioning of an enzyme. Co-factors are usually loosely associated with the enzyme.

Conditioned medium: Culture medium that has supported growth of cells. Such medium will contain a range of cell-derived molecules (e.g. amino acids, growth substances) that may enhance the growth of a subsequent batch of cells.

Cycloheximide: A reversible inhibitor of eukaryotic protein synthesis isolated from *Streptomyces griseus*. It acts by inhibiting peptidyl transferase.

Cytochrome P-450: An electron transferring ferroprotein family, that participate in a range of cellular redox reactions.

Cytokinin: Plant hormone traditionally considered to promote cell division e.g. zeatin (natural) and 6BA (synthetic).

2,4-D: 2,4-dichlorophenoxyacetic acid. An auxin.

Dicotyledon: One of the two subgroups of angiosperms, characterised by the possession of two cotyledons. See monocotyledon.

Differentiated: Cell showing characteristics different from meristem or germ cells. The differentiated state is usually stable. In plant cell culture, the term differentiation is often taken to mean shoot or root formation.

Diterpenoid: One of a class of C_{20} compounds derived from geranyl geranyl pyrosphosphate.

Doubling time (^{t}d): The time required to complete one cell division cycle.

Elicitor: When used in the context of phytoalexin formation, an elicitor is a substance, often an oligosaccharide of cell wall origin, which evokes the production of phytoalexin pathotoxins.

Elite organism: An indiviudal plant (or organism in general) that possesses desirable characteristics that are not shared by the majority of that species or variety.

Epigenetic variation: Inheritable variation in the characteristics of organisms that is not due to changes in the primary sequence of nucleotides in the DNA. An example is the switching off of genes by base methylation.

Explant: That part of a plant that is excised for use in the initiation of a cell or tissue culture.

Exponential phase: The phase of batch growth of cells during which there is a continuous division of daughter cells such that growth parameters increase in an exponential manner.

Fed batch system: A culture system in a bioreactor in which frequent additions of medium or carbon source are made without removal of culture.

Fluid mosaic model: The model of membrane structure put forward by Singer and Nicholson. It proposes that membrane components can diffuse in the plane of the membrane.

Fluidized bed: A fluidized bed reactor is one containing particles or cellular aggregates which are kept in suspension by passage of liquid or gas through the bed.

Germplasm: Genetic material in a form from which the whole organism can be recovered, i.e. cells capable of regeneration. In the case of plants germplasm may be seeds, shoot cultures or regenerable cell cultures.

Glandular hair: A differentiated multicellular structure often found in leaves, consisting of a uni- or multicellular stalk and a unicellular head. The secretion of oil beneath the cuticle may raise it, as in *Mentha* and *Cannabis*.

Glycosidase: The generic name for an enzyme that cleaves a glycosidic bond, i.e. an enzyme that removes sugars from a glycoconjugate.

Gold-immune (immunogold) labelling: A technique for localising antigens in electron microscope (em) sections. The sections are treated with the appropriate antibody linked to gold particles. Gold, being electron dense, is easily observed under em and marks the position of the antigen.

Haemocytometer: A microscope slide ruled in grids, that is used for counting cells.

Indole 3-acetic acid (IAA): A natural plant auxin.

Infra-red spectroscopy: Analytical method based upon monitoring the electromagnetic spectrum of a complex molecule between about 7.5×10^{-5} to 4.3×10^{-2} cm wavelength range. This includes photons produced or absorbed during rotational and vibrational transitions.

Ion exchange resin: An electrically-charged material that will bind ions of the opposite charge. Bound ions can be released by changing the pH or ionic strength. Ion-exchange resins are extensively used for biological purifications.

Isoprene (unit): A 5 carbon unit shown to be the building block of all isoprenoid structures ranging in complexity from monoterpenes to rubber. The 'active isoprene unit' in metabolism is isopentenyl pyrophosphate.

Lag phase: The initial phase of batch growth of cells characterised by little change in growth parameters. Lag phases result from the time required to synthesise the enzymes required for nutrient utilization, efflux and equilibration of metabolites into the new growth medium, and a stress response (e.g. osmotic) to the inoculation procedure.

Laticifer: A cell or tube, often arising from the fusion of longitudinal series of cells, containing latex; latex is a fluid with a milky appearance arising from suspended small particles (e.g. essential oils) in a liquid dispersion medium.

Liquid/gas decoupling: Liquid/gas decoupling is the separation of gas, usually as bubbles from the culture medium at the top of a bioreactor. The space required for this depends on the gas flow rates and foaming.

Mass spectrometry (ms): Analytical method based upon the separation of charged particles according to their mass-to-charge ratios. A plot or mass spectrum is produced of the fragmentation patterns of complex molecules after their bombardment with electrons (electron impact mass spectrometry).

Matrix (cell wall): The hemicellulose, pectic and protein compounds that cement together the cellulose microfibrils in plant cell primary walls.

Meristem: A group of small, actively-dividing cells the derivatives of which differentiate into the tissues of the plant body. Root and shoot apical meristems are highly organised structures. The latter give rise to precise patterns of buds and leaves.

Meristematic: Having meristem-like characteristics.

Mesophyll cells: The green, photosynthetic cells between the two epidermal layers in leaves. The cells are relatively undifferentiated and often retain the ability to divide readily.

Methotrexate: Amethopterin; a folic acid analogue that inhibits the enzyme dihydrofolate reductase.

Microsomes: An operational name given to small membrane vesicles appearing when cells are homogenised. They are fragments of the various membrane systems of the cell e.g. endoplasmic reticulum, Golgi bodies and plasma membranes.

Minimum inoculation density: Cell density below which a culture will not survive and grow when transferred to fresh medium.

Monoclonal antibody: A homogeneous preparation of antibody i.e. one in which all antibodies have the same specificity (antigen binding site). Such antibodies are produced by cloning individual antibody-producing cells.

Monocotyledon: One of the two subgroups of angiosperms characterised by having a single cotyledon. This group of plants contains the agriculturally important grasses and palms. From a cell culture standpoint, monocots are relatively intractable and generally inaccessible to *Agrobacterium* based transformation (at present).

Monoterpenoid: Isoprenoid chemical derived from geranyl pyrophosphate (C_{10}). There are many different structures, most of which constitute important components of volatile oils.

NAA: 1-naphthylacetic acid.

Nuclear magnetic resonance (nmr): Method for studying the interaction of an atomic nucleus possessing an odd mass number and odd proton number with the nuclear environment. A characteristic profile can be produced for individual chemicals.

Packed bed: A packed bed reactor is a vessel containing particles or cell aggregates which, restricted in their movement, form a solid bed through which fluid can flow.

Pectinase: An enzyme that hydrolyses pectin, a heterogeneous polysaccharide mainly found in the middle lamella between adjacent plant cell walls. Incubation of tissue with pectinase can produce individual, walled cells.

Phytoalexin: a non-specific antibiotic substance produced by plant cells in response to infection by pathogens, or environmental stress or injury. Phytoalexins are often phenolic molecules.

Plagiotropic: A growth movement (tropism) that proceeds at an angle to the stimulus.

Plasmolysis: The removal of water from a cell, with attendant shrinkage, resulting from hyperosmotic extracellular conditions.

Ploidy: The extent of duplication of the basic chromosome set (designated n) of a cell e.g. haploid (n) diploid (2n) polyploid (4n, 8n etc).

Polyclonal antibodies: A preparation containing antibodies with different specificities. Such antibodies are produced when an antigen is injected into an animal. Even if a pure antigen is used, different antibody-producing cells will form antibodies to different domains on the antigen surface.

Polyembryonic: Containing more than one embryo. Usually applied to a fertilized ovule or seed.

Polygenic trait: A phenotypic character that is specified by many individual genes.

Promoter (e.g. CaMv 35S promoter): The region of a gene to which binds the RNA polymerase enzyme. Certain strong promoters (e.g. CaMV 35S) are favoured in gene manipulation experiments because they give a high expression of the gene in question.

Pyridine alkaloid: Alkaloid structure based upon a pyridine ring (N-substitution in benzene ring).

Racemic mixtures: A mixture of the D- and L- optical isomers of a molecule. They are produced when an asymmetric carbon atom passes through a symmetrical intermediate state such that subsequent isomerisation is a random process.

Recycle system: A system whereby cells are removed from medium leaving a bioreactor and returned to the bioreactor.

Regeneration: Development of cultured tissue, cells or protoplasts towards the whole plant. It can refer to wall resynthesis or colony formation in protoplasts, or organ or whole plantlet formation in cells or tissue.

Regioselective: Recognising the spatial orientation of a functional group in relation to its immediate molecular environment.

Ri-DNA: DNA from *Agrobacterium rhizogenes* that is transferred into the host cell. Transfer of wild-type Ri DNA results in "hairy root" disease.

Secondary metabolites: A metabolite that is not part of the primary metabolic network of the cells. Primary metabolites are components of fundamental biochemical pathways (e.g. glycolysis) that are present in all cells. Secondary metabolites only appear in cells that specialise in some way e.g. in defence.

Semi-continuous culture: The growth of cells in a bioreactor, in which medium is periodically replaced to prolong the culture, but a balance between cell growth and removal is not reached; the cells are not in a steady state.

Sesquiterpenoid: Isoprenoid chemical derived from farnesyl pyrophosphate (C_{15}). These can be linear, monocylic or bicyclic in structure, and many are known to be 'stress' compounds.

Single pass plug flow: This describes the flow of medium or substrate through a reactor or bioreactor which is unmixed. Bioreactors supporting this type of flow are often tubular.

Somatic embryo: An embryo-like structure originating from a somatic (non-germ) cell.

Somatic hybrid: A hybrid resulting from the fusion of non-germ cells.

Sparging: Sparging is the introduction of gas to a reactor.

Specific growth rate: Estimation of growth, derived from a logarithmic plot of biomass increase versus time. It is a constant defined as the rate of growth per unit of organism concentrate.

Spiroketal: Polyacetylenic compound, derived from long chain alcohols and characteristic of some plants of the Compositae family.

Stationary phase: The phase of batch growth of cells during which there is no change in the growth parameter being measured. A stationary phase is usually due to depletion of nutrients from the growth medium.

Steric hindrance: The physical interference between atoms in a molecule. This interference restricts the number of configurations that the molecule can adopt.

Sucrose density gradient centrifugation: A separation technique for isolating particles (cells or macromolecules) based on their characteristic densities. The particles reach an equilibrium position at which the sucrose solution has the same density.

Suspension culture: A culture of cells distributed throughout liquid growth medium.

2,4,5-T: 2,4,5-trichlorophenoxyacetic acid. A synthetic auxin.

Tannin: A polymeric phenol having the characteristic property of forming a precipitate with protein. Tannins fall into two classes; hydrolysable and condensed.

Totipotent (cells): Cells having the complete genetic potential (ie the full set of genes) of the whole plant ie. they can in principle regenerate back into plants and express the plants, complete range of metabolism.

Transgenic: Possessing genes from a different organism.

Transposable elements: Sections of DNA that can reposition themselves in a chromosome. The element usually codes for a specific nuclease that excises the DNA.

Tropane alkaloid: Alkaloid structure including a nitrogen – containing 5 membered heterocyclic ring and derived from hygrine or ornithine.

Undifferentiated (tissue cultures): Showing no morphological cellular specialisation e.g. vascular elements, cell alignments,, root or shoot formation etc. The term cannot be used in an absolute sense because biochemical specialisation may easily go unnoticed.

Vacuolation: The formation of visible vacuoles in a plant cell. The process usually results from the fusion and enlargement of numerous small vacuoles and is often accompanied by cell elongation.

Variant/mutant: An organism showing a stable phenotypic deviation from the wild type. Such an organism can only be classed as a true mutant if an equivalent change in the primary structure of the genome can be demonstrated.

Vegetative (cells): All cells other than germ cells.

Xenobiotic: Pollutants, for example pesticide chemicals such as polychlorinated and polycyclic hydrocarbons released into the environment.

Zygotic embryo: An embryo arising from a fertilized ovule as in the normal life cycle of the plant.

Index